Ben Stacy Jerrik (Ed.)

Naum Meiman

Ben Stacy Jerrik (Ed.)

# Naum Meiman

## Soviet Dissidents

**Part Press**

# Contents

## Articles



## References

# Naum Meiman

| Naum Meiman | |
|---|---|
| **Born** | May 12, 1912Baranovichi, now Belarus, then Russian empire |
| **Died** | 3.31.2001Tel-Aviv, Israel |
| **Residence** | Germany |
| **Nationality** | USSR<br>Israel |
| **Fields** | Mathematician |
| **Institutions** | Kazan State University, University of Kharkiv, Institute for Physical Problems, Institute for Theoretical and Experimental Physics, Tel Aviv University |
| **Alma mater** | Kazan State University |
| **Doctoral advisor** | Nikolai Chebotaryov |
| **Notable awards** | USSR State Prize |

**Naum S. Meiman** (Russian: Нау́м Ната́нович Ме́йман) (1911, Baranovichi, Minsk region — 2001, Tel-Aviv) was a Soviet mathematician, and dissident. [1] His is known for his work in complex analysis, partial differential equations, and mathematical physics, as well as for his dissident activity, in particular, for being a member of the Moscow Helsinki Group.

## Life

Meiman is to the right in the group of pupils of Nikolai Chebotaryov (to the left)

In 1932 he graduated Kazan State University as an extern. In 1937 being only 26 years old, he submitted his Ph.D. under the supervision of Nikolai Chebotaryov, and was immediately awarded the degree Doktor nauk. In 1939 he became a full professor in Kazan State University.[1]

He worked for two years in the Mathematics Institute at the University of Kharkiv, where he became friends with Lev Landau with whom he collaborated for many years. After the Second World War, he went to Moscow and worked in Institute for Physical Problems, where he was a head of the mathematics lab. Then he worked in the Institute for Theoretical and Experimental Physics. In 1953, he was awarded a Stalin prize for his work in theoretical physics. He made important contributions into the development of nuclear weapons in the USSR.

Group of Soviet dissidents, first from left: Naum Meiman, the first from right: Andrei Sakharov

Starting in 1968, Meiman became active in politics and signed several letters of protest against political trials in USSR.[1]

In 1971, he retired and applied for permission to emigrate to Israel. Denied on grounds of knowing state secrets, he soon became a refusenik. Gradually he became more active in politics, and was a member of Moscow Helsinki Group beginning in 1977. Later he became deputy chairman and the last active free member, writing hundreds of the group documents. He also participated in a Refusenik scientific seminar. He was permanently under surveillance by the KGB, who also bugged his telephone and searched his home.

Meiman also struggled for the right of his wife Inna Meiman-Kitrossky to go to USA for medical treatment since she had been diagnosed with cancer. After several years of struggle, she was allowed to go US and died in February, 1987 in Georgetown (Washington, D.C.).[2] [3] [4] Meiman was not allowed to attend her funeral in Washington D.C.[5] [6]

In 1988 Meiman was finally allowed to emigrate to Israel, where he became a professor emeritus in Tel Aviv University. In 1992, in Tel-Aviv, there was a conference in his honor dedicated to his 80th birthday. Meiman died there in 2001.[1] He has a daughter, Olga, who lives in the USA.

## References

[1]  Anosov, D.V.; Ginzburg, V.L.; Zhizhchenko, A.B.; Monastyrskii, M.I.; Novikov, S.P.; Sinai, Ya.G.; Solov'ev, M.A. (2002). "Naum Meiman. Obituary" (http://iopscience.iop.org/0036-0279/57/2/M05/). *Russ. Math. Surv.* **57** (2): 399–405. doi:10.1070/RM2002v057n02ABEH000499. .

[2]  Paul, Lisa (2011). *Swimming in the Daylight: An American Student, a Soviet-Jewish Dissident, and the Gift of Hope* (http://www.goodreads. com/book/show/10295308-swimming-in-the-daylight). Skyhorse Publishing. ISBN 978-1616082031. .

[3]  Lisa Paul presents her book (http://www.youtube.com/watch?v=sseV26HziPc)

[4]  Dr. Leonid Stonov. "Book review" (http://www.angelfire.com/sc3/soviet_jews_exodus/English/Chronicle_s/Stonov_Review.shtml). Association «Remember and save». . Retrieved 2011-02-23.

[5]  INNA MEIMAN, EMIGRE, DIES AT 53 (http://query.nytimes.com/gst/fullpage. html?res=9B0DE5DE173EF933A25751C0A961948260&sec=health&spon=&partner=permalink&exprod=permalink), NY Times

[6]  OLD AND ALONE, SOVIET DISSIDENT LOOKS TO EXIT (http://query.nytimes.com/gst/fullpage. html?res=9B0DEFDE153EF935A15751C0A961948260&scp=5&sq=Meiman), NY Times

## External links

- SOVIET HUMAN RIGHTS BATTLE: ONLY ISOLATED VOICES REMAIN (http://query.nytimes.com/gst/ fullpage.html?sec=health&res=9404E3DB1338F93AA15754C0A963948260&scp=4&sq=Meiman), NY Times
- INNA MEIMAN, EMIGRE, DIES AT 53 (http://query.nytimes.com/gst/fullpage. html?res=9B0DE5DE173EF933A25751C0A961948260&sec=health&spon=&partner=permalink& exprod=permalink), NY Times
- OLD AND ALONE, SOVIET DISSIDENT LOOKS TO EXIT (http://query.nytimes.com/gst/fullpage. html?res=9B0DEFDE153EF935A15751C0A961948260&scp=5&sq=Meiman), NY Times

## Online journals

- Московская Хельсинкская группа (Moscow Helsinki group). Мейман Наум Натанович (Naum Meiman) (http:/ /www.mhg.ru/history/1B3270D)^Russian:
- Naum Meiman. List of works (http://www.mathnet.ru/links/b28770486a7b12c2e5d662d364524672/ rm_499_refsN/rm_499_refsN.html)

# Soviet dissidents

**Soviet dissidents** were citizens of the Soviet Union who disagreed with the policies and actions of their government and actively protested against these actions through either violent or non-violent means. Through such protests, Soviet dissidents incurred harassment, persecution, imprisonment or death by the KGB, or other Soviet government agencies.

From the mid-1970s, the term was first used in the Western propaganda [1] and subsequently, with derision, by the Soviet media. Human rights activists in the USSR then adopted this term.

While dissent with Soviet policies and persecution for this dissent existed since the times of the October Revolution and the establishment of the Soviet power, the term is most commonly applied to the dissidents of the post-Stalin era.

## See also

- Alexsandr Solzhenitsyn
- Andrei Sakharov
- Chronicle of Current Events (samizdat)
- Committee on Human Rights in the USSR
- Gulag
- Moscow Helsinki Group
- Samizdat
- Yelena Bonner

## References

[1] Доклад: Диссиденты (http://www.solzhenicyn.ru/modules/sections/index_op_viewarticle_artid_257.html) by Michel Aucouturier

## Further reading

- Roy Aleksandrovich Medvedev, Piero Ostellino (1980) *On Soviet Dissent*, ISBN 0231048122
- Robert Horvath (2005) *The Legacy of Soviet Dissent: Dissidents, Democratisation and Radical Nationalism in Russia*, ISBN 0415333202
- Dissenters, Soviet Union (http://books.google.com/books?q=+subject:"Dissenters;+Soviet+Union.") on Google Books

# Baranovichi

<table>
<tr><td colspan="2" align="center">Baranavichy<br>Баранавічы<br>Барановичи</td></tr>
<tr><td colspan="2"></td></tr>
<tr><td colspan="2" align="center"><br>Seal</td></tr>
<tr><td colspan="2" align="center">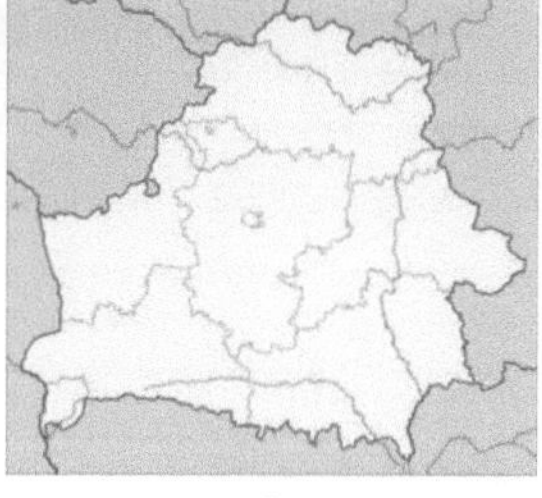<br>Baranavichy<br><br>Location in Belarus</td></tr>
<tr><td colspan="2" align="center">Coordinates: 53°08′N 26°01′E</td></tr>
<tr><td>Country</td><td>Belarus</td></tr>
<tr><td>Voblast</td><td>Brest Voblast</td></tr>
<tr><td>Raion</td><td>Baranavichy Raion</td></tr>
<tr><td>Mentioned</td><td>1706</td></tr>
<tr><td colspan="2">Population (1995)</td></tr>
<tr><td>• Total</td><td>173,000</td></tr>
<tr><td>Time zone</td><td>EET (UTC+2)</td></tr>
</table>

| • **Summer (DST)** | EEST (UTC+3) |
| --- | --- |
| **Postal code** | 225320 |
| **Area code(s)** | +375 (0)163 |
| **Website** | www.baranovichi.by [1] |

**Baranovichi** English pronunciation: /bəˈrɑːnəvɪtʃiː/ ( ◀ listen) (Belarusian: Баранавічы Belarusian pronunciation: [baˈranavʲitsɨ], **Baranavichy**; Russian: Бара́новичи, Polish: *Baranowicze*, Yiddish: שטיוואנאראב, *Baranovich*), is a city in the Brest Province of western Belarus with a population (as of 1995) of 173,000. It is a significant railway junction and home to a state university.

## History

### Overview

The village of Baranowicze, as it was originally called, was first mentioned in 1706 as a private property of a Polish family named Rozwadowski. In the late 18th century, in the effect of the Partitions of Poland, the town became part of the Russian Empire. In the 1870s the locality became an important railway junction, on the crossing of the Warsaw-Moscow and Vilnia-Lviv lines, and was renamed Baranovichi after the Baranovich family.[2] Soon the village started to grow, and by 1883 it became a town of almost 2,000 inhabitants. By 1897 the population of the town had risen to 4,600 inhabitants (ca. 50% Jews). At the beginning of the First World War it was chosen as the location of the Stavka, which was removed to Mogilev in August 1915.

Former Baranovichi Law Institute, Now a constituent part of Baranovichi State University

Baranovichi. Fountain at Central Square

### Recent history

During the Polish-Soviet war it was reclaimed by Poland. In 1919 it received city rights. In 1921 Baranowicze had over 11,000 inhabitants (67% Jews, with the rest being mostly Belarusians, Poles and Russians). Soon the town started to grow and became an important centre of trade and commerce for the area. The town's Orthodox cathedral was built in the Neoclassical style in 1924-31; it was decorated with mosaics that had survived the demolition of the Alexander Nevsky Cathedral, Warsaw. The town was also an important military garrison, with one KOP Cavalry Brigade, 20th Infantry

Ballistic missile on display in Baranovichi

Division and the Nowogródzka Cavalry Brigade stationed there. Because of the fast growth of local industry, in 1938 a local branch of the Polish Radio was opened there. In 1939 Baranovichi had almost 30,000 inhabitants and was the biggest and the most important city in the Nowogródek Voivodship.

Soon after the beginning of World War II the town was occupied by the Soviet Union. The local Jewish population of 9,000 was joined by approximately 3,000 Jewish refugees from the Polish areas occupied by Germany. After the start of Operation Barbarossa the town was seized by the Wehrmacht on June 25, 1941. In August of the same year a ghetto was created in the town, with more than 12,000 Jews kept in tragic conditions in six buildings at the outskirts. Between March 4 and December 14, 1942, the entire Jewish population of the ghetto was sent to various German concentration camps and killed in gas chambers. Only approximately 250 survived the war.[3]

The town was occupied by the Red Army since July 6, 1944. Significant part of the Polish population of the town had been expelled to Siberia and Kazakhstan. After the World War II the town became part of the Soviet Union and the Byelorussian SSR and started to be referred to under its Russian name of Baranovichi. In this time an intensive industrialization took place. In 1991 it became part of independent Belarus.

## Twin towns

- **Biała Podlaska** (Poland)
- **Ferrara** (Italy)
- **Gdynia** (Poland)
- **Heinola** (Finland)
- **Jelgava** (Latvia)
- **Karlovo** (Bulgaria)
- **Kineshma** (Russia)
- **Mytishchi** (Russia)
- **Novovolynsk** (Ukraine)
- **Poltava** (Ukraine)
- **Stockerau** (Austria)
- **Zhodzina** (Belarus)

## See also

- FC Baranovichi
- Polish Radio Baranowicze

## References

[1] http://www.baranovichi.by

[2] E. M. Pospelov, *Geograficheskie nazvaniya mira* (Moscow, 1998), p. 56.

[3] Baranovichi. Holocaust (http://jhrgbelarus.org/Heritage_Holocaust.php?pid=&lang=en&city_id=1&type=3)

## External links

- BRI.BY - City site of Baranovichi (http://www.bri.by)
- BARANOVICHI.BY - CITY PORTAL (http://baranovichi.by)
- INTEX-PRESS online - latest news of Baranovichi region (http://www.intex-press.by)
- Football in Baranovichi (http://www.fcbaranovichi.com)
- Your city. Your portal. Website of the City of Baranovichi (http://baranovichi.by)
- Baranovichi informational portal (http://baranavichy.info)
- Baranavichy in history (http://baranavichy.at.tut.by)
- History of Baranovichi (http://www.kresy.co.uk/baranowicze.html)
- Baranowicze Radio Station (http://www.historiaradia.neostrada.pl/Stacja nadawcza Baranowicze.html)
- Pre-war photos of Baranovichi (http://www.ketrzyn.mm.pl/~wwmkiewicz/ws/baranowicze/)

- Historic images of Baranovichi (http://www.bfcollection.net/cities/belarus/baranovichi/baranovichi_01.
  html)
- Modern views of Baranovichi (http://globus.tut.by/baranovich/index.htm)
- Photos on Radzima.org (http://www.radzima.org/pub/miesta.php?lang=en&miesta_id1=brbabara)
- Public Transport in Baranavichy (http://bartrans.net/uk.html)
- Baranovichi University Photos (http://ruscakursu.info/baranovichi/)
- Baranovichi. Synagogues (http://jhrgbelarus.org/Heritage_Synagogues.php?pid=&lang=en&city_id=1&
  type=1)

# Mathematician

A **mathematician** is a person whose primary area of study is the field of mathematics. Mathematicians are concerned with quantity, structure, space, and change.

Some scientists who research other fields, such as theoretical physics, are also considered mathematicians if their research provides insights into mathematics. Conversely, some mathematicians may provide insights into other fields of research—these people are known as applied mathematicians.

## Education

Mathematicians usually cover breadth of topics within mathematics in their undergraduate education, and then proceed to specialize in topics of their own choice at the graduate level. In some universities, a qualifying exam serves to test both the breadth and depth of a student's understanding of mathematics; the students who pass are permitted to work on a doctoral dissertation. There are notable cases where mathematicians have failed to reflect their ability in their university

Archimedes was among the greatest mathematicians of antiquity.

education, but have nevertheless become remarkable mathematicians. Fermat, for example, is known for having been "Prince of Amateurs", because of his extraordinary achievements with little formal mathematics training.[1]

## Motivation

Mathematicians do research in fields such as logic, set theory, category theory, abstract algebra, number theory, analysis, geometry, topology, dynamical systems, combinatorics, game theory, information theory, numerical analysis, optimization, computation, probability and statistics. These fields comprise both pure mathematics and applied mathematics and establish links between the two. Some fields, such as the theory of dynamical systems, or game theory, are classified as applied mathematics due to the relationships they possess with physics, economics and the other sciences. Whether probability theory and statistics are of theoretical nature, applied nature, or both, is quite controversial among mathematicians. Other branches of mathematics, however, such as logic, number theory, category theory or set theory are accepted to be a part of pure mathematics, although they do indeed find applications in other sciences (predominantly computer science and physics). Likewise, analysis, geometry and topology, although considered pure mathematics, do find applications in theoretical physics - string theory, for instance.

Although it is true that mathematics finds diverse applications in many areas of research, a mathematician does not determine the value of an idea by the diversity of its applications. Mathematics is interesting in its own right, and a majority of mathematicians investigate the diversity of structures studied in *mathematics itself*. Furthermore, a mathematician is not someone who merely manipulates formulas, numbers or equations—the diversity of mathematics allows for research concerning how concepts in one area of mathematics can be used in other areas too. For instance, if one graphs a set of solutions of an equation in some higher dimensional space, he may ask about the geometric properties of the graph. Thus one can understand equations by a pure understanding of abstract topology or geometry—this idea is of importance in algebraic geometry. Similarly, a mathematician does not restrict his study of numbers to the integers; rather he considers more abstract structures such as rings, and in particular number rings in the context of algebraic number theory. This exemplifies the abstract nature of mathematics and how it is not restricted to questions one may ask in daily life.

In a different direction, mathematicians ask questions about space and transformations, but which are not restricted to geometric figures such as squares and circles. For instance, an active area of research within the field of differential topology concerns itself with the ways in which one can "smooth" higher dimensional figures. In fact, whether one can smooth certain higher dimensional spheres was, until recently, an open problem — known as the smooth Poincaré conjecture. Another aspect of mathematics, set-theoretic topology and point-set topology, concerns objects of a different nature from objects in our universe, or in a higher dimensional analogue of our universe. These objects behave in a rather strange manner under deformations, and the properties they possess are completely different from those of objects in our universe. For instance, the "distance" between two points on such an object, may depend on the order in which you consider the pair of points. This is quite different from ordinary life, in which it is accepted that the straight line distance from person A to person B is the same as that between person B and person A.

Isaac Newton was a pioneering figure in the development of mathematical physics.

Leonhard Euler is widely considered one of the greatest mathematicians.

Another aspect of mathematics, often referred to as "foundational mathematics", consists of the fields of logic and set theory. Here, various ideas regarding the ways in which one can prove certain claims are explored. This theory is far more complex than it seems, in that the truth of a claim depends on the context in which the claim is made, unlike basic ideas in daily life where truth is absolute. In fact, although some claims may be true, it is impossible to prove or disprove them in rather natural contexts.

Category theory, another field within "foundational mathematics", is rooted on the abstract axiomatization of the definition of a "class of mathematical structures", referred to as a "category". A category intuitively consists of a collection of objects, and defined relationships between them. While these objects may be anything (such as "tables" or "chairs"), mathematicians are usually interested in particular, more abstract, classes of such objects. In any case, it is the *relationships between these objects*, and *not the actual objects* which are predominantly studied.

Carl Friedrich Gauss was the foremost mathematician of the early 19th century.

## Differences with scientists

Mathematics differs from natural sciences in that physical theories in the sciences are tested by experiments, while mathematical statements are supported by proofs that may be verified objectively. If a certain statement is believed to be true by mathematicians (typically because special cases have been confirmed to some degree) but has neither been proved nor disproved, it is called a *conjecture*, as opposed to the ultimate goal: a *theorem* that has been proved. Physical theories may be expected to change whenever new information about our physical world is discovered. Mathematics changes in a different way: new ideas do not falsify old ones but rather are used to *generalize* what was known before to capture a broader range of phenomena. For instance, calculus (in one variable) generalizes to multivariable calculus, which generalizes to analysis on manifolds. The development of algebraic geometry from its classical to modern forms is a particularly striking example of the way an area of mathematics can change radically in its viewpoint without making what was proved before in any way

Henri Poincaré is considered to be the last mathematician to excel in every field of the mathematics of his time.

incorrect. While a theorem, once proved, is true forever, our understanding of what the theorem *really means* gains in profundity as the mathematics around the theorem grows. A mathematician feels that a theorem is better understood when it can be extended to apply in a broader setting than previously known. For instance, Fermat's little theorem for the nonzero integers modulo a prime generalizes to Euler's theorem for the invertible numbers modulo any nonzero integer, which generalizes to Lagrange's theorem for finite groups.

## Women in mathematics

While the majority of mathematicians are male, there have been some demographic changes since World War II. Some prominent female mathematicians are Hypatia of Alexandria (ca. 400 AD), Ada Lovelace (1815–1852), Maria Gaetana Agnesi (1718–1799), Emmy Noether (1882–1935), Sophie Germain (1776–1831), Sofia Kovalevskaya (1850–1891), Alicia Boole Stott (1860–1940), Rózsa Péter (1905–1977), Julia Robinson (1919–1985), Olga Taussky-Todd (1906–1995), Émilie du Châtelet (1706–1749), Mary Cartwright (1900–1998), and Olga Ladyzhenskaya (1922–2004).

The Association for Women in Mathematics is a professional society whose purpose is "to encourage women and girls to study and to have active careers in the mathematical sciences, and to promote equal opportunity and the equal treatment of women and girls in the mathematical sciences." The American Mathematical Society and other mathematical societies offer several prizes aimed at increasing the representation of women and minorities in the future of mathematics.

David Hilbert was one of the most influential mathematicians of the late 19th and early 20th centuries.

Emmy Noether is possibly the most influential female mathematician to date.

## Prizes in mathematics

There is no Nobel Prize in mathematics, though sometimes mathematicians have won the Nobel Prize in a different field, such as economics. Prominent prizes in mathematics include the Abel Prize, the Chern Medal, the Fields Medal, the Gauss Prize, the Nemmers Prize, the Balzan Prize, the Crafoord Prize, the Shaw Prize, the Wolf Prize, the Schock Prize, and the Nevanlinna Prize.

## Quotations about mathematicians

The following are quotations about mathematicians, or by mathematicians.

> *A mathematician is a device for turning coffee into theorems.*
>
> —Attributed to both Alfréd Rényi[2] and Paul Erdős

> *Die Mathematiker sind eine Art Franzosen; redet man mit ihnen, so übersetzen sie es in ihre Sprache, und dann ist es alsobald ganz etwas anderes.* (Mathematicians are [like] a sort of Frenchmen; if you talk to them, they translate it into their own language, and then it is immediately something quite different.)
>
> —Johann Wolfgang von Goethe

> *Each generation has its few great mathematicians...and [the others'] research harms no one.*
>
> —Alfred W. Adler (1930- ), "Mathematics and Creativity"[3]

> *In short, I never yet encountered the mere mathematician who could be trusted out of equal roots, or one who did not clandestinely hold it as a point of his faith that x squared + px was absolutely and unconditionally equal to q. Say to one of these gentlemen, by way of experiment, if you please, that you believe occasions may occur where x squared + px is not altogether equal to q, and, having made him understand what you mean, get out of his reach as speedily as convenient, for, beyond doubt, he will endeavor to knock you down.*
>
> —Edgar Allan Poe, *The purloined letter*

> *A mathematician, like a painter or poet, is a maker of patterns. If his patterns are more permanent than theirs, it is because they are made with ideas.*
>
> —G. H. Hardy, *A Mathematician's Apology*

> *Some of you may have met mathematicians and wondered how they got that way.*
>
> —Tom Lehrer

> *It is impossible to be a mathematician without being a poet in soul.*
>
> —Sofia Kovalevskaya

> *There are two ways to do great mathematics. The first is to be smarter than everybody else. The second way is to be stupider than everybody else—but persistent.*
>
> —Raoul Bott

## See also

Some notable mathematicians include Archimedes of Syracuse, Leonhard Euler, Carl Gauss, Johann Bernoulli, Jacob Bernoulli, Aryabhatta, Brahmagupta,Bhaskara II,Nilakantha Somayaji,Omar Khayyám, Muhammad ibn Mūsā al-Khwārizmī, Bernhard Riemann, Gottfried Leibniz, Andrey Kolmogorov, Euclid of Alexandria, Jules Henri Poincaré, Srinivasa Ramanujan, Alexander Grothendieck, David Hilbert, Alan Turing, von Neumann, Kurt Gödel, Joseph-Louis Lagrange, Georg Cantor, William Rowan Hamilton, Carl Jacobi, Évariste Galois, Nikolay Lobachevsky, Rene Descartes, Joseph Fourier, Pierre-Simon Laplace, Robert Lee Moore, Alonzo Church, Nikolay Bogolyubov and Pierre de Fermat.

• List of amateur mathematicians

- List of female mathematicians
- List of mathematicians
- Mathematical joke
- Men of mathematics (book)
- A Mathematician's Apology

## Notes

[1] http://mathsforeurope.digibel.be/pierredefermat.html
[2] Biography of Alfréd Rényi (http://www-history.mcs.st-andrews.ac.uk/Biographies/Renyi.html)
[3] Alfred Adler, "Mathematics and Creativity," *The New Yorker*, 1972, reprinted in Timothy Ferris, ed., *The World Treasury of Physics, Astronomy, and Mathematics*, Back Bay Books, reprint, June 30, 1993, p, 435.

## References

- *A Mathematician's Apology*, by G. H. Hardy. Memoir, with foreword by C. P. Snow.
    - Reprint edition, Cambridge University Press, 1992; ISBN 0-521-42706-1
    - First edition, 1940
- Paul Halmos. *I Want to Be a Mathematician*. Springer-Verlag 1985.
- Dunham, William. *The Mathematical Universe*. John Wiley 1994.

## External links

- Occupational Outlook: Mathematicians (http://stats.bls.gov/oco/ocos043.htm). Information on the occupation of mathematician from the US Department of Labor.
- Sloan Career Cornerstone Center: Careers in Mathematics (http://www.careercornerstone.org/math/math. htm). Although US-centric, a useful resource for anyone interested in a career as a mathematician. Learn what mathematicians do on a daily basis, where they work, how much they earn, and more.
- The MacTutor History of Mathematics archive (http://www-history.mcs.st-and.ac.uk/history/index0.html). A comprehensive list of detailed biographies.
- The Mathematics Genealogy Project (http://genealogy.math.ndsu.nodak.edu/). Allows to follow the succession of thesis advisors for most mathematicians, living or dead.
- Weisstein, Eric W., " Unsolved Problems (http://mathworld.wolfram.com/UnsolvedProblems.html)" from MathWorld.
- Middle School Mathematician Project (http://valure.wiki.ccsd.edu/) Short biographies of select mathematicians assembled by middle school students.
- Career Information for Students of Math and Aspiring Mathematicians (http://www.mathmajor.org/careers) from MathMajor (http://mathmajor.org)

# Complex analysis

**Complex analysis**, traditionally known as the **theory of functions of a complex variable**, is the branch of mathematical analysis that investigates functions of complex numbers. It is useful in many branches of mathematics, including number theory and applied mathematics; as well as in physics, including hydrodynamics, thermodynamics, and electrical engineering.

Murray R. Spiegel described complex analysis as "one of the most beautiful as well as useful branches of Mathematics".

Complex analysis is particularly concerned with the analytic functions of complex variables (or, more generally, meromorphic functions). Because the separate real and imaginary parts of any analytic function must satisfy Laplace's equation, complex analysis is widely applicable to two-dimensional problems in physics.

Plot of the function $f(x) = (x^2 - 1)(x - 2 - i)^2 / (x^2 + 2 + 2i)$. The hue represents the function argument, while the brightness represents the magnitude.

## History

Complex analysis is one of the classical branches in mathematics with roots in the 19th century and just prior. Important names are Euler, Gauss, Riemann, Cauchy, Weierstrass, and many more in the 20th century. Complex analysis, in particular the theory of conformal mappings, has many physical applications and is also used throughout analytic number theory. In modern times, it has become very popular through a new boost from complex dynamics and the pictures of fractals produced by iterating holomorphic functions. Another important application of complex analysis is in string theory which studies conformal invariants in quantum field theory.

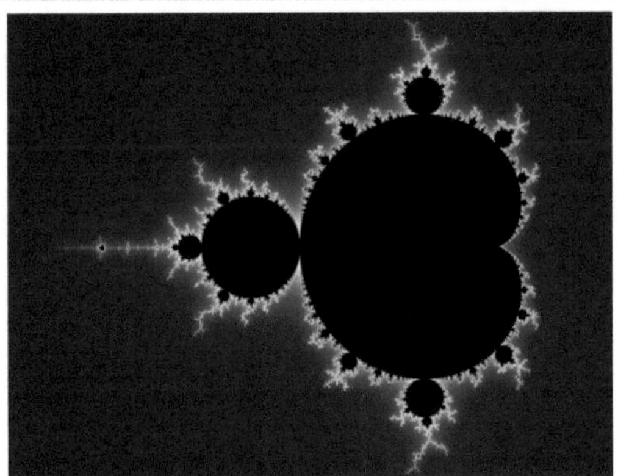

The Mandelbrot set, a **fractal**.

## Complex functions

A complex function is one in which the independent variable and the dependent variable are both complex numbers. More precisely, a complex function is a function whose domain and range are subsets of the complex plane.

For any complex function, both the independent variable and the dependent variable may be separated into real and imaginary parts:

and

where and are real-valued functions.

In other words, the components of the function $f(z)$,

and

can be interpreted as real-valued functions of the two real variables, $x$ and $y$.

The basic concepts of complex analysis are often introduced by extending the elementary real functions (e.g., exponentials, logarithms, and trigonometric functions) into the complex domain.

## Holomorphic functions

Holomorphic functions are complex functions defined on an open subset of the complex plane that are differentiable. Complex differentiability has much stronger consequences than usual (real) differentiability. For instance, holomorphic functions are infinitely differentiable, whereas some real differentiable functions are not. Most elementary functions, including the exponential function, the trigonometric functions, and all polynomial functions, are holomorphic.

*See also*: analytic function, holomorphic sheaf and vector bundles.

## Major results

One central tool in complex analysis is the line integral. The integral around a closed path of a function that is holomorphic everywhere inside the area bounded by the closed path is always zero; this is the Cauchy integral theorem. The values of a holomorphic function inside a disk can be computed by a certain path integral on the disk's boundary (Cauchy's integral formula). Path integrals in the complex plane are often used to determine complicated real integrals, and here the theory of residues among others is useful (see methods of contour integration). If a function has a *pole* or *singularity* at some point, that is, at that point where its values "blow up" and have no finite boundary, then one can compute the function's residue at that pole. These residues can be used to compute path integrals involving the function; this is the content of the powerful residue theorem. The remarkable behavior of holomorphic functions near essential singularities is described by Picard's Theorem. Functions that have only poles but no essential singularities are called meromorphic. Laurent series are similar to Taylor series but can be used to study the behavior of functions near singularities.

A bounded function that is holomorphic in the entire complex plane must be constant; this is Liouville's theorem. It can be used to provide a natural and short proof for the fundamental theorem of algebra which states that the field of complex numbers is algebraically closed.

If a function is holomorphic throughout a simply connected domain then its values are fully determined by its values on any smaller subdomain. The function on the larger domain is said to be analytically continued from its values on the smaller domain. This allows the extension of the definition of functions, such as the Riemann zeta function, which are initially defined in terms of infinite sums that converge only on limited domains to almost the entire complex plane. Sometimes, as in the case of the natural logarithm, it is impossible to analytically continue a holomorphic function to a non-simply connected domain in the complex plane but it is possible to extend it to a holomorphic function on a closely related surface known as a Riemann surface.

All this refers to complex analysis in one variable. There is also a very rich theory of complex analysis in more than one complex dimension in which the analytic properties such as power series expansion carry over whereas most of the geometric properties of holomorphic functions in one complex dimension (such as conformality) do not carry over. The Riemann mapping theorem about the conformal relationship of certain domains in the complex plane, which may be the most important result in the one-dimensional theory, fails dramatically in higher dimensions.

## See also

- Complex dynamics
- List of complex analysis topics
- Real analysis
- Runge's theorem
- Several complex variables

## Notes

## References


- Ahlfors.,*Complex Analysis*(McGraw-Hill).
- C.Caratheodory, *Theory of Functions of a Complex Variable* (Chelsea,New York). [2 volumes.]
- Needham T., *Visual Complex Analysis* (Oxford, 1997).
- Henrici P., *Applied and Computational Complex Analysis* (Wiley). [Three volumes: 1974, 1977, 1986.]
- Kreyszig, E., *Advanced Engineering Mathematics, 9 ed.*, Ch.13-18 (Wiley, 2006).
- A.I.Markushevich.,*Theory of Functions of a Complex Variable*(Prentice-Hall, 1965).[Three volumes.]
- Scheidemann, V., *Introduction to complex analysis in several variables* (Birkhauser, 2005)
- Shaw, W.T., *Complex Analysis with Mathematica* (Cambridge, 2006).
- Spiegel, Murray R. *Theory and Problems of Complex Variables - with an introduction to Conformal Mapping and its applications* (McGraw-Hill, 1964).
- Marsden & Hoffman, *Basic complex analysis* (Freeman, 1999).


## External links

- Complex Analysis -- textbook by George Cain (http://www.math.gatech.edu/~cain/winter99/complex.html)
- Complex analysis course web site (http://www.ima.umn.edu/~arnold/502.s97/) by Douglas N. Arnold
- Example problems in complex analysis (http://www.exampleproblems.com/wiki/index.php/Complex_Variables)
- A collection of links to programs for visualizing complex functions (and related) (http://www.usfca.edu/vca/websites.html)
- Complex Analysis Project by John H. Mathews (http://math.fullerton.edu/mathews/complex.html)
- Wolfram Research's MathWorld Complex Analysis Page (http://mathworld.wolfram.com/ComplexAnalysis.html)
- Application of Complex Functions in 2D Digital Image Transformation (http://vadim-kataev.livejournal.com/135060.html)
- Complex Visualizer - Java applet for visualizing arbitrary complex functions (http://www.saunalahti.fi/mattpaa/complex/complex.html)
- JavaScript complex function graphing tool (http://www.fortwain.com/complex.html)
- Earliest Known Uses of Some of the Words of Mathematics: Calculus & Analysis (http://www.economics.soton.ac.uk/staff/aldrich/Calculus and Analysis Earliest Uses.htm)

# Partial differential equation

In mathematics, **partial differential equations** (PDE) are a type of differential equation, i.e., a relation involving an unknown function (or functions) of several independent variables and their partial derivatives with respect to those variables. PDEs are used to formulate, and thus aid the solution of problems involving functions of several variables.

PDEs are for example used to describe the propagation of sound or heat, electrostatics, electrodynamics, fluid flow, and elasticity. These seemingly distinct physical phenomena can be formalized identically (in terms of PDEs), which shows that they are governed by the same underlying dynamic. PDEs find their generalization in stochastic partial differential equations. Just as ordinary differential equations often model dynamical systems, partial differential equations often model multidimensional systems.

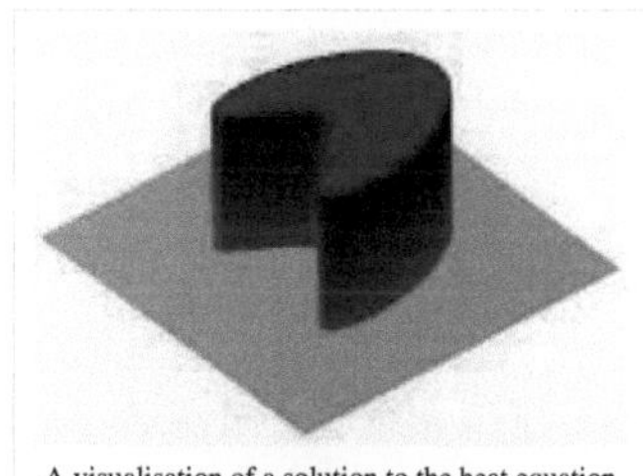

A visualisation of a solution to the heat equation on a two dimensional plane

## Introduction

A partial differential equation (PDE) for the function is an equation of the form

If $F$ is a linear function of $u$ and its derivatives, then the PDE is called linear. Common examples of linear PDEs include the heat equation, the wave equation and Laplace's equation.

A relatively simple PDE is

This relation implies that the function $u(x,y)$ is independent of $x$. Hence the general solution of this equation is

where $f$ is an arbitrary function of $y$. The analogous ordinary differential equation is

which has the solution

where $c$ is any constant value (independent of $x$). These two examples illustrate that general solutions of ordinary differential equations (ODEs) involve arbitrary constants, but solutions of PDEs involve arbitrary functions. A solution of a PDE is generally not unique; additional conditions must generally be specified on the boundary of the region where the solution is defined. For instance, in the simple example above, the function can be determined if is specified on the line .

## Existence and uniqueness

Although the issue of existence and uniqueness of solutions of ordinary differential equations has a very satisfactory answer with the Picard–Lindelöf theorem, that is far from the case for partial differential equations. There is a general theorem (the Cauchy–Kowalevski theorem) that states that the Cauchy problem for any partial differential equation whose coefficients are analytic in the unknown function and its derivatives, has a locally unique analytic solution. Although this result might appear to settle the existence and uniqueness of solutions, there are examples of linear partial differential equations whose coefficients have derivatives of all orders (which are nevertheless not analytic) but which have no solutions at all: see Lewy (1957). Even if the solution of a partial differential equation exists and is unique, it may nevertheless have undesirable properties. The mathematical study of these questions is usually in the more powerful context of weak solutions.

An example of pathological behavior is the sequence of Cauchy problems (depending upon $n$) for the Laplace equation

with boundary conditions

where $n$ is an integer. The derivative of $u$ with respect to $y$ approaches 0 uniformly in $x$ as $n$ increases, but the solution is

This solution approaches infinity if $nx$ is not an integer multiple of $\pi$ for any non-zero value of $y$. The Cauchy problem for the Laplace equation is called *ill-posed* or *not well posed*, since the solution does not depend continuously upon the data of the problem. Such ill-posed problems are not usually satisfactory for physical applications.

## Notation

In PDEs, it is common to denote partial derivatives using subscripts. That is:

Especially in (mathematical) physics, one often prefers the use of del (which in cartesian coordinates is written ) for spatial derivatives and a dot for time derivatives. For example, the wave equation (described below) can be written as

(physics notation),

or

(math notation),

where $\Delta$ is the Laplace operator.

## Examples

### Heat equation in one space dimension

The equation for conduction of heat in one dimension for a homogeneous body has the form

where $u(t,x)$ is temperature, and $\alpha$ is a positive constant that describes the rate of diffusion. The Cauchy problem for this equation consists in specifying , where $f(x)$ is an arbitrary function.

General solutions of the heat equation can be found by the method of separation of variables. Some examples appear in the heat equation article. They are examples of Fourier series for periodic $f$ and Fourier transforms for non-periodic $f$. Using the Fourier transform, a general solution of the heat equation has the form

where $F$ is an arbitrary function. To satisfy the initial condition, $F$ is given by the Fourier transform of $f$, that is

If $f$ represents a very small but intense source of heat, then the preceding integral can be approximated by the delta distribution, multiplied by the strength of the source. For a source whose strength is normalized to 1, the result is

and the resulting solution of the heat equation is

This is a Gaussian integral. It may be evaluated to obtain

This result corresponds to the normal probability density for $x$ with mean 0 and variance $2\alpha t$. The heat equation and similar diffusion equations are useful tools to study random phenomena.

### Wave equation in one spatial dimension

The wave equation is an equation for an unknown function $u(t, x)$ of the form

Here $u$ might describe the displacement of a stretched string from equilibrium, or the difference in air pressure in a tube, or the magnitude of an electromagnetic field in a tube, and $c$ is a number that corresponds to the velocity of the wave. The Cauchy problem for this equation consists in prescribing the initial displacement and velocity of a string or other medium:

where $f$ and $g$ are arbitrary given functions. The solution of this problem is given by d'Alembert's formula:

This formula implies that the solution at $(t,x)$ depends only upon the data on the segment of the initial line that is cut out by the characteristic curves that are drawn backwards from that point. These curves correspond to signals that propagate with velocity $c$ forward and backward. Conversely, the influence of the data at any given point on the initial line propagates with the finite velocity $c$: there is no effect outside a triangle through that point whose sides are characteristic curves. This behavior is very different from the solution for the heat equation, where the effect of a point source appears (with small amplitude) instantaneously at every point in space. The solution given above is also valid if $t$ is negative, and the explicit formula shows that the solution depends smoothly upon the data: both the forward and backward Cauchy problems for the wave equation are well-posed.

## Generalised heat-like equation in one space dimension

Where heat-like equation means equations of the form:

Where is a Sturm-Liouville operator (However it should be noted this operator may in fact be of the form where

$w(x)$ is the weighting function with respect to which the eigenfunctions of are orthogonal) in the x coordinate.

Subject to the boundary conditions:

Then:

If:

Where

## Spherical waves

Spherical waves are waves whose amplitude depends only upon the radial distance $r$ from a central point source. For such waves, the three-dimensional wave equation takes the form

This is equivalent to

and hence the quantity $ru$ satisfies the one-dimensional wave equation. Therefore a general solution for spherical waves has the form

where $F$ and $G$ are completely arbitrary functions. Radiation from an antenna corresponds to the case where $G$ is identically zero. Thus the wave form transmitted from an antenna has no distortion in time: the only distorting factor is $1/r$. This feature of undistorted propagation of waves is not present if there are two spatial dimensions.

## Laplace equation in two dimensions

The Laplace equation for an unknown function of two variables $\varphi$ has the form

Solutions of Laplace's equation are called harmonic functions.

### Connection with holomorphic functions

Solutions of the Laplace equation in two dimensions are intimately connected with analytic functions of a complex variable (a.k.a. holomorphic functions): the real and imaginary parts of any analytic function are **conjugate harmonic** functions: they both satisfy the Laplace equation, and their gradients are orthogonal. If $f=u+iv$, then the Cauchy–Riemann equations state that

and it follows that

Conversely, given any harmonic function in two dimensions, it is the real part of an analytic function, at least locally. Details are given in Laplace equation.

**A typical boundary value problem**

A typical problem for Laplace's equation is to find a solution that satisfies arbitrary values on the boundary of a domain. For example, we may seek a harmonic function that takes on the values $u(\theta)$ on a circle of radius one. The solution was given by Poisson:

Petrovsky (1967, p. 248) shows how this formula can be obtained by summing a Fourier series for $\varphi$. If $r<1$, the derivatives of $\varphi$ may be computed by differentiating under the integral sign, and one can verify that $\varphi$ is analytic, even if $u$ is continuous but not necessarily differentiable. This behavior is typical for solutions of elliptic partial differential equations: the solutions may be much more smooth than the boundary data. This is in contrast to solutions of the wave equation, and more general hyperbolic partial differential equations, which typically have no more derivatives than the data.

# Euler–Tricomi equation

The Euler–Tricomi equation is used in the investigation of transonic flow.

# Advection equation

The advection equation describes the transport of a conserved scalar in a velocity field . It is:

If the velocity field is solenoidal (that is, ), then the equation may be simplified to

In the one-dimensional case where is not constant and is equal to , the equation is referred to as Burgers' equation.

# Ginzburg–Landau equation

The Ginzburg–Landau equation is used in modelling superconductivity. It is

where and are constants and is the imaginary unit.

# The Dym equation

The Dym equation is named for Harry Dym and occurs in the study of solitons. It is

# Initial-boundary value problems

Further information: Examples of boundary value problems

Many problems of mathematical physics are formulated as initial-boundary value problems.

**Vibrating string**

If the string is stretched between two points where $x=0$ and $x=L$ and $u$ denotes the amplitude of the displacement of the string, then $u$ satisfies the one-dimensional wave equation in the region where $0<x<L$ and $t$ is unlimited. Since the string is tied down at the ends, $u$ must also satisfy the boundary conditions

as well as the initial conditions

The method of separation of variables for the wave equation

leads to solutions of the form

where

where the constant $k$ must be determined. The boundary conditions then imply that $X$ is a multiple of $\sin kx$, and $k$ must have the form where $n$ is an integer. Each term in the sum corresponds to a mode of vibration of the string. The mode with $n=1$ is called the fundamental mode, and the frequencies of the other modes are all multiples of this frequency. They form the overtone series of the string, and they are the basis for musical acoustics. The initial conditions may then be satisfied by representing $f$ and $g$ as infinite sums of these modes. Wind instruments typically correspond to vibrations of an air column with one end open and one end closed. The corresponding boundary conditions are

The method of separation of variables can also be applied in this case, and it leads to a series of odd overtones.

The general problem of this type is solved in Sturm–Liouville theory.

### Vibrating membrane

If a membrane is stretched over a curve $C$ that forms the boundary of a domain $D$ in the plane, its vibrations are governed by the wave equation

if $t>0$ and $(x,y)$ is in $D$. The boundary condition is if is on . The method of separation of variables leads to the form

which in turn must satisfy

The latter equation is called the Helmholtz Equation. The constant $k$ must be determined to allow a non-trivial $v$ to satisfy the boundary condition on $C$. Such values of $k^2$ are called the eigenvalues of the Laplacian in $D$, and the associated solutions are the eigenfunctions of the Laplacian in $D$. The Sturm–Liouville theory may be extended to this elliptic eigenvalue problem (Jost, 2002).

## Other examples

The Schrödinger equation is a PDE at the heart of non-relativistic quantum mechanics. In the WKB approximation it is the Hamilton–Jacobi equation.

Except for the Dym equation and the Ginzburg–Landau equation, the above equations are **linear** in the sense that they can be written in the form $Au = f$ for a given linear operator $A$ and a given function $f$. Other important non-linear equations include the Navier–Stokes equations describing the flow of fluids, and Einstein's field equations of general relativity.

Also see the list of non-linear partial differential equations.

# Classification

Some linear, second-order partial differential equations can be classified as parabolic, hyperbolic or elliptic. Others such as the Euler–Tricomi equation have different types in different regions. The classification provides a guide to appropriate initial and boundary conditions, and to smoothness of the solutions.

## Equations of second order

Assuming , the general second-order PDE in two independent variables has the form

where the coefficients $A$, $B$, $C$ etc. may depend upon $x$ and $y$. If over a region of the xy plane, the PDE is second-order in that region. This form is analogous to the equation for a conic section:

More precisely, replacing by and likewise for other variables (formally this is done by a Fourier transform), converts a constant-coefficient PDE into a polynomial of the same degree, with the top degree (a homogeneous polynomial, here a quadratic form) being most significant for the classification.

Just as one classifies conic sections and quadratic forms into parabolic, hyperbolic, and elliptic based on the discriminant , the same can be done for a second-order PDE at a given point. However, the discriminant in a PDE is given by due to the convention of the $xy$ term being $2B$ rather than $B$; formally, the discriminant (of the associated quadratic form) is with the factor of 4 dropped for simplicity.

1.  : solutions of elliptic PDEs are as smooth as the coefficients allow, within the interior of the region where the equation and solutions are defined. For example, solutions of Laplace's equation are analytic within the domain where they are defined, but solutions may assume boundary values that are not smooth. The motion of a fluid at subsonic speeds can be approximated with elliptic PDEs, and the Euler–Tricomi equation is elliptic where $x<0$.
2.  : equations that are parabolic at every point can be transformed into a form analogous to the heat equation by a change of independent variables. Solutions smooth out as the transformed time variable increases. The Euler–Tricomi equation has parabolic type on the line where $x=0$.
3.  : hyperbolic equations retain any discontinuities of functions or derivatives in the initial data. An example is the wave equation. The motion of a fluid at supersonic speeds can be approximated with hyperbolic PDEs, and the Euler–Tricomi equation is hyperbolic where $x>0$.

If there are $n$ independent variables $x_1, x_2, ..., x_n$, a general linear partial differential equation of second order has the form

The classification depends upon the signature of the eigenvalues of the coefficient matrix.

1.  Elliptic: The eigenvalues are all positive or all negative.
2.  Parabolic : The eigenvalues are all positive or all negative, save one that is zero.
3.  Hyperbolic: There is only one negative eigenvalue and all the rest are positive, or there is only one positive eigenvalue and all the rest are negative.
4.  Ultrahyperbolic: There is more than one positive eigenvalue and more than one negative eigenvalue, and there are no zero eigenvalues. There is only limited theory for ultrahyperbolic equations (Courant and Hilbert, 1962).

## Systems of first-order equations and characteristic surfaces

The classification of partial differential equations can be extended to systems of first-order equations, where the unknown $u$ is now a vector with $m$ components, and the coefficient matrices are $m$ by $m$ matrices for . The partial differential equation takes the form

where the coefficient matrices $A_\nu$ and the vector $B$ may depend upon $x$ and $u$. If a hypersurface $S$ is given in the implicit form

where $\varphi$ has a non-zero gradient, then $S$ is a **characteristic surface** for the operator $L$ at a given point if the characteristic form vanishes:

The geometric interpretation of this condition is as follows: if data for $u$ are prescribed on the surface $S$, then it may be possible to determine the normal derivative of $u$ on $S$ from the differential equation. If the data on $S$ and the differential equation determine the normal derivative of $u$ on $S$, then $S$ is non-characteristic. If the data on $S$ and the differential equation *do not* determine the normal derivative of $u$ on $S$, then the surface is **characteristic**, and the differential equation restricts the data on $S$: the differential equation is *internal* to $S$.

1.  A first-order system $Lu=0$ is *elliptic* if no surface is characteristic for $L$: the values of $u$ on $S$ and the differential equation always determine the normal derivative of $u$ on $S$.
2.  A first-order system is *hyperbolic* at a point if there is a **space-like** surface $S$ with normal $\xi$ at that point. This means that, given any non-trivial vector $\eta$ orthogonal to $\xi$, and a scalar multiplier $\lambda$, the equation

has $m$ real roots $\lambda_1, \lambda_2, ..., \lambda_m$. The system is **strictly hyperbolic** if these roots are always distinct. The geometrical interpretation of this condition is as follows: the characteristic form $Q(\zeta)=0$ defines a cone (the normal cone) with homogeneous coordinates $\zeta$. In the hyperbolic case, this cone has $m$ sheets, and the axis $\zeta = \lambda \xi$ runs inside these sheets: it does not intersect any of them. But when displaced from the origin by $\eta$, this axis intersects every sheet. In the elliptic case, the normal cone has no real sheets.

## Equations of mixed type

If a PDE has coefficients that are not constant, it is possible that it will not belong to any of these categories but rather be of **mixed type**. A simple but important example is the Euler–Tricomi equation

which is called **elliptic-hyperbolic** because it is elliptic in the region $x < 0$, hyperbolic in the region $x > 0$, and degenerate parabolic on the line $x = 0$.

## Infinite-order PDEs in quantum mechanics

Weyl quantization in phase space leads to quantum Hamilton's equations for trajectories of quantum particles. Those equations are infinite-order PDEs. However, in the semiclassical expansion one has a finite system of ODEs at any fixed order of . The equation of evolution of the Wigner function is infinite-order PDE also. The quantum trajectories are quantum characteristics with the use of which one can calculate the evolution of the Wigner function.

# Analytical methods to solve PDEs

## Separation of variables

In the method of separation of variables, one reduces a PDE to a PDE in fewer variables, which is an ODE if in one variable – these are in turn easier to solve.

This is possible for simple PDEs, which are called separable partial differential equations, and the domain is generally a rectangle (a product of intervals). Separable PDEs correspond to diagonal matrices – thinking of "the value for fixed $x$" as a coordinate, each coordinate can be understood separately.

This generalizes to the method of characteristics, and is also used in integral transforms.

## Method of characteristics

In special cases, one can find characteristic curves on which the equation reduces to an ODE – changing coordinates in the domain to straighten these curves allows separation of variables, and is called the method of characteristics.

More generally, one may find characteristic surfaces.

## Integral transform

An integral transform may transform the PDE to a simpler one, in particular a separable PDE. This corresponds to diagonalizing an operator.

An important example of this is Fourier analysis, which diagonalizes the heat equation using the eigenbasis of sinusoidal waves.

If the domain is finite or periodic, an infinite sum of solutions such as a Fourier series is appropriate, but an integral of solutions such as a Fourier integral is generally required for infinite domains. The solution for a point source for the heat equation given above is an example for use of a Fourier integral.

## Change of variables

Often a PDE can be reduced to a simpler form with a known solution by a suitable change of variables. For example the Black–Scholes PDE

is reducible to the heat equation

by the change of variables (for complete details see Solution of the Black Scholes Equation [1])

## Fundamental solution

Inhomogeneous equations can often be solved (for constant coefficient PDEs, always be solved) by finding the fundamental solution (the solution for a point source), then taking the convolution with the boundary conditions to get the solution.

This is analogous in signal processing to understanding a filter by its impulse response.

## Superposition principle

Because any superposition of solutions of a linear, homogeneous PDE is again a solution, the particular solutions may then be combined to obtain more general solutions.

## Methods for non-linear equations

*See also the list of nonlinear partial differential equations.*

There are no generally applicable methods to solve non-linear PDEs. Still, existence and uniqueness results (such as the Cauchy–Kowalevski theorem) are often possible, as are proofs of important qualitative and quantitative properties of solutions (getting these results is a major part of analysis). Computational solution to the nonlinear PDEs, the split-step method, exist for specific equations like nonlinear Schrödinger equation.

Nevertheless, some techniques can be used for several types of equations. The h-principle is the most powerful method to solve underdetermined equations. The Riquier–Janet theory is an effective method for obtaining information about many analytic overdetermined systems.

The method of characteristics (Similarity Transformation method) can be used in some very special cases to solve partial differential equations.

In some cases, a PDE can be solved via perturbation analysis in which the solution is considered to be a correction to an equation with a known solution. Alternatives are numerical analysis techniques from simple finite difference schemes to the more mature multigrid and finite element methods. Many interesting problems in science and engineering are solved in this way using computers, sometimes high performance supercomputers.

## Lie group method

From 1870 Sophus Lie's work put the theory of differential equations on a more satisfactory foundation. He showed that the integration theories of the older mathematicians can, by the introduction of what are now called Lie groups, be referred to a common source; and that ordinary differential equations which admit the same infinitesimal transformations present comparable difficulties of integration. He also emphasized the subject of transformations of contact.

A general approach to solve PDE's uses the symmetry property of differential equations, the continuous infinitesimal transformations of solutions to solutions (Lie theory). Continuous group theory, Lie algebras and differential geometry are used to understand the structure of linear and nonlinear partial differential equations for generating integrable equations, to find its Lax pairs, recursion operators, Bäcklund transform and finally finding exact analytic solutions to the PDE.

Symmetry methods have been recognized to study differential equations arising in mathematics, physics, engineering, and many other disciplines.

## Semianalytical method

The adomian decomposition method, the Lyapunov artificial small parameter method and He's homotopy perturbation method are all special cases of the more general homotopy analysis method. There are series expansion method but there are independent of small physical parameters compared to thee well known perturbation theory.

# Numerical methods to solve PDEs

The three most widely used numerical methods to solve PDEs are the finite element method (FEM), finite volume methods (FVM) and finite difference methods (FDM). The FEM has a prominent position among these methods and especially its exceptionally efficient higher-order version hp-FEM. Other versions of FEM include the generalized finite element method (GFEM), extended finite element method (XFEM), spectral finite element method (SFEM), meshfree finite element method, discontinuous Galerkin finite element method (DGFEM), etc.

## Finite element method

The finite element method (FEM) (its practical application often known as finite element analysis (FEA)) is a numerical technique for finding approximate solutions of partial differential equations (PDE) as well as of integral equations. The solution approach is based either on eliminating the differential equation completely (steady state problems), or rendering the PDE into an approximating system of ordinary differential equations, which are then numerically integrated using standard techniques such as Euler's method, Runge-Kutta, etc.

## Finite difference method

Finite-difference methods are numerical methods for approximating the solutions to differential equations using finite difference equations to approximate derivatives.

## Finite volume method

Similar to the finite difference method or finite element method, values are calculated at discrete places on a meshed geometry. "Finite volume" refers to the small volume surrounding each node point on a mesh. In the finite volume method, volume integrals in a partial differential equation that contain a divergence term are converted to surface integrals, using the divergence theorem. These terms are then evaluated as fluxes at the surfaces of each finite volume. Because the flux entering a given volume is identical to that leaving the adjacent volume, these methods are conservative.

# See also

- Boundary value problem
- Difference equation
- Laplace transform applied to differential equations
- List of dynamical systems and differential equations topics
- Matrix differential equation
- Ordinary differential equation
- Separation of variables
- Stochastic partial differential equations
- Numerical partial differential equations
- Stochastic processes and boundary value problems

- Dirichlet boundary condition
- Neumann boundary condition
- Robin boundary condition
- waves

## References


- Adomian, G. (1994). *Solving Frontier problems of Physics: The decomposition method*. Kluwer Academic Publishers.
- Courant, R. & Hilbert, D. (1962), *Methods of Mathematical Physics*, **II**, New York: Wiley-Interscience.
- Evans, L. C. (1998), *Partial Differential Equations*, Providence: American Mathematical Society, ISBN 0821807722.
- Ibragimov, Nail H (1993), *CRC Handbook of Lie Group Analysis of Differential Equations Vol. 1-3*, Providence: CRC-Press, ISBN 0849344883.
- John, F. (1982), *Partial Differential Equations* (4th ed.), New York: Springer-Verlag, ISBN 0387906096.
- Jost, J. (2002), *Partial Differential Equations*, New York: Springer-Verlag, ISBN 0387954287.
- Lewy, Hans (1957), "An example of a smooth linear partial differential equation without solution", *Annals of Mathematics, 2nd Series* **66** (1): 155–158.
- Liao, S.J. (2003), *Beyond Perturbation: Introduction to the Homotopy Analysis Method*, Boca Raton: Chapman & Hall/ CRC Press, ISBN 158488407X
- Olver, P.J. (1995), *Equivalence, Invariants and Symmetry*, Cambridge Press.
- Petrovskii, I. G. (1967), *Partial Differential Equations*, Philadelphia: W. B. Saunders Co..
- Pinchover, Y. & Rubinstein, J. (2005), *An Introduction to Partial Differential Equations*, New York: Cambridge University Press, ISBN 0521848865.
- Polyanin, A. D. (2002), *Handbook of Linear Partial Differential Equations for Engineers and Scientists*, Boca Raton: Chapman & Hall/CRC Press, ISBN 1584882999.
- Polyanin, A. D. & Zaitsev, V. F. (2004), *Handbook of Nonlinear Partial Differential Equations*, Boca Raton: Chapman & Hall/CRC Press, ISBN 1584883553.
- Polyanin, A. D.; Zaitsev, V. F. & Moussiaux, A. (2002), *Handbook of First Order Partial Differential Equations*, London: Taylor & Francis, ISBN 041527267X.
- Solin, P. (2005), *Partial Differential Equations and the Finite Element Method*, Hoboken, NJ: J. Wiley & Sons, ISBN 0471720704.
- Solin, P.; Segeth, K. & Dolezel, I. (2003), *Higher-Order Finite Element Methods*, Boca Raton: Chapman & Hall/CRC Press, ISBN 158488438X.
- Stephani, H. (1989), *Differential Equations: Their Solution Using Symmetries. Edited by M. MacCallum*, Cambridge University Press.
- Wazwaz, Abdul-Majid (2009). *Partial Differential Equations and Solitary Waves Theory*. Higher Education Press. ISBN 9058093697.
- Zwillinger, D. (1997), *Handbook of Differential Equations* (3rd ed.), Boston: Academic Press, ISBN 0127843957.

## External links

- Partial Differential Equations: Exact Solutions [2] at EqWorld: The World of Mathematical Equations.
- Partial Differential Equations: Index [3] at EqWorld: The World of Mathematical Equations.
- Partial Differential Equations: Methods [4] at EqWorld: The World of Mathematical Equations.
- Example problems with solutions [5] at exampleproblems.com
- Partial Differential Equations [6] at mathworld.wolfram.com
- Dispersive PDE Wiki [7]
- NEQwiki, the nonlinear equations encyclopedia [8]

## References

[1] http://web.archive.org/web/20080411030405/http://www.math.unl.edu/~sdunbar1/Teaching/MathematicalFinance/Lessons/
    BlackScholes/Solution/solution.shtml

[2] http://eqworld.ipmnet.ru/en/pde-en.htm

[3] http://eqworld.ipmnet.ru/en/solutions/eqindex/eqindex-pde.htm

[4] http://eqworld.ipmnet.ru/en/methods/meth-pde.htm

[5] http://www.exampleproblems.com/wiki/index.php?title=Partial_Differential_Equations

[6] http://mathworld.wolfram.com/PartialDifferentialEquation.html

[7] http://tosio.math.toronto.edu/wiki/index.php/Main_Page

[8] http://www.primat.mephi.ru/wiki/

# Moscow Helsinki Group

The **Moscow Helsinki Group** (also known as the **Moscow Helsinki Watch Group**, Russian: Московская Хельсинкская группа) is an influential human rights monitoring non-governmental organization, originally established in what was then the Soviet Union;[1] it still operates in Russia.

It was founded in 1976 to monitor the Soviet Union's compliance with the recently-signed Helsinki Final Act of 1975, which included clauses calling for the recognition of universal human rights. Its pioneering efforts inspired the formation of similar groups in other Warsaw Pact countries and support groups in the West. In Czechoslovakia, Charter 77 was founded in January 1977; members of that group would later play key roles in the overthrow of the communist dictatorship in Czechoslovakia. In Poland, a Helsinki Watch Group was founded in September 1979. Eventually, the collection of Helsinki monitoring groups inspired by the Moscow Helsinki Group formed the International Helsinki Federation.

## Details

Helsinki monitoring efforts began in the Soviet Union shortly after the publication of the Helsinki Final Act in Soviet newspapers.

On May 12, 1976, physicist Yuri Orlov announced the formation of the "Public Group to Promote Fulfillment of the Helsinki Accords in the USSR" (Общественная группа содействия выполнению хельсинкских соглашений в СССР, Московская группа "Хельсинки") at a press-conference held at the apartment of Andrei Sakharov. The newly inaugurated NGO was meant to monitor Soviet compliance with the Helsinki Final Act. The eleven founders of the group also included Lyudmila Alexeyeva, Mikhail Bernshtam, Yelena Bonner, Alexander Ginzburg, Pyotr Grigorenko, Alexander Korchak, Malva Landa, Anatoly Marchenko, Gregory Rosenstein, Vitaly Rubin, and Anatoly Shcharansky. Ten other people, including Sofia Kalistratova, Naum Meiman, Yuri Mniukh, Victor Nekipelov, Tatiana Osipova, Felix Serebrov, Vladimir Slepak, Leonard Ternovsky, and Yuri Yarym-Agaev joined the Group later.

The group's goal was to uphold the government of the Soviet Union to implement the commitment to human rights it had made in the Helsinki documents. Their group's legal validity was based on the provision in the Helsinki Final Act, Principle VII, which establishes the rights of individuals to know and act upon their rights and duties.

The Soviet authorities responded with severe repression of the group's members over the following three years. They used tactics that included arrests and imprisonment, internal exile, confinement to psychiatric hospitals, and forced emigration. On 18 October 1976, 13 Jewish *refuseniks* came to the Presidium of the Supreme Soviet to petition for explanations of denials of their right to emigrate from the U.S.S.R., as affirmed under the Helsinki Final Act. Failing to receive any answer, they assembled in the reception room of the Presidium on the following day. After a few hours of waiting, they were seized by the agents of militia, taken outside of the city limits and beaten. Two of them were kept in police custody. In the next week, following an unsuccessful meeting between the activists' leaders and the Soviet Minister of Internal Affairs, General Nikolay Shchelokov, these abuses of law inspired several mass demonstrations in the Soviet capital. On Monday, October 25, 22 activists, including Mark Azbel, Felix Kandel, Alexander Lerner, Ida Nudel, Anatoly Shcharansky, Vladimir Slepak, and Michael Zeleny, were arrested in Moscow on their way to the next demonstration. They were convicted of hooliganism and incarcerated in the detention center Beryozka and other penitentiaries in and around Moscow. An unrelated party, artist Victor Motko, arrested in Dzerzhinsky Square on the account of wearing a woolly black beard, was detained along with the protesters in recognition of his prior attempts to emigrate from the U.S.S.R. These events were covered by several British and American journalists including David K. Shipler, Craig R. Whitney, and Christopher S. Wren. The October demonstrations and arrests coincided with the end of the 1976 United States presidential election. On October 25, U.S. Presidential candidate Jimmy Carter expressed his support of the protesters in a telegram sent to Scharansky, and urged the Soviet authorities to release them. (See Léopold Unger, Christian Jelen, *Le grand retour*, A. Michel 1977; Феликс Кандель, *Зона отдыха, или Пятнадцать суток на размышление*, Типография Ольшанский Лтд, Иерусалим, 1979; Феликс Кандель, *Врата исхода нашего: Девять страниц истории*, Effect Publications, Tel-Aviv, 1980.) On 9 November 1976, a week after Carter won the Presidential election, the Soviet authorities released all but two of the previously arrested protesters. Several more were subsequently rearrested and incarcerated or exiled to Siberia.

On 1 June 1978, *refuseniks* Vladimir and Maria Slepak stood on the eighth story balcony of their apartment building. By then they had been denied permission to emigrate for over 8 years. Vladimir displayed a banner that read "Let us go to our son in Israel". His wife Maria held a banner that read "Visa for my son". Fellow *refusenik* and Helsinki activist Ida Nudel held a similar display on the balcony of her own apartmemt. They were all arrested and charged with malicious hooliganism in violation of Article 206.2 of the Penal Code of the Soviet Union. The Helsinki Group protested their arrests in circulars dated 5 and 15 June of that year. ([2]) Vladimir Slepak and Ida Nudel were convicted of all charges. They served 5 and 4 years in Siberian exile. ([3], [4])

## The Working Commission to Investigate the Use of Psychiatry For Political Purposes

In January 1977, Alexandr Podrabinek along with a 47 year-old self-educated worker Feliks Serebrov, a 30 year-old computer programmer Vyacheslav Bakhmin and Irina Kuplun established the Working Commission to Investigate the Use of Psychiatry for Political Purposes.[5] :148 The Commission was formally linked to[5] :148 and constituted as an offshoot of the Moscow Helsinki Group.[6] The commission was composed of five open members and several anonymous ones, including a few psychiatrists who, at great danger to themselves, conducted their own independent examinations of cases of alleged psychiatric abuse.[7] The leader of the commission was Alexandr Podrabinek who published a book *Punitive Medicine*[7] containing a 'white list' of two hundred of prisoners of conscience in Soviet mental hospitals and a 'black list' of over one hundred medical staff and doctors who took part in committing people to psychiatric facilities for political reasons.[8] :15

The psychiatric consultants to the Commission were Dr Alexander Voloshanovich and Dr Anatoly Koryagin.[9] :153 The task stated by the Commission was not primarily to diagnose persons or to declare people who sought help mentally ill or mentally healthy.[5] :150[10] :26 However, in some instances individuals who came for help to the Commission were examined by a psychiatrist who provided help to the Commission and made a precise diagnosis of their mental condition.[5] :150[10] :26 At first it was psychiatrist Aleksandr Voloshanovich from the Moscow suburb of Dolgoprudny, who made these diagnoses.[5] :150 But when he had been compelled to emigrate on 7 February 1980,[11] his work was continued by the Kharkov psychiatrist Anatoly Koryagin.[5] :150 Koryagin's contribution was to examine former and potential victims of political abuse of psychiatry by writing psychiatric diagnoses in which he deduced that the individual was not suffering from any mental disease.[5] :179 Those reports were employed as a means of defense: if the individual was picked up again and committed to mental hospital, the Commission had vindication that the hospitalization served non-medical purposes.[5] :179 Also some foreign psychiatrists including the Swedish psychiatrist Harald Blomberg and British psychiatrist Gery Low-Beer helped in examining former or potential victims of psychiatric abuse.[5] :150 The Commission used those reports in its work and publicly referred to them when it was essential.[5] :150

The commission gathered as much information as possible of victims of psychiatric terror in the Soviet Union and published this information in their *Information Bulletins*.[12] :45 For the four years of its existence, the Commission published more than 1,500 pages of documentation including 22 *Information Bulletins* in which over 400 cases of the political abuse of psychiatry were documented in great detail.[5] :148 Summaries of the *Information Bulletins* were published in the key samizdat publication, the *Chronicle of Current Events*.[5] :148 The *Information Bulletins* were sent to the Soviet officials, with request to verify the data and notify the Commission if mistakes were found, and to the West, where human rights defenders used them in the course of their campaigns.[5] :148 The *Information Bulletins* were also used to provide the dissident movement with information about Western protests against the political abuse.[5] :148 Peter Reddaway said that after he had studied official documents in the Soviet archives, including minutes from meetings of the Politburo of the Central Committee of the Communist Party of the Soviet Union, it became evident to him that Soviet officials at high levels paid close attention to foreign responses to these cases, and if someone was discharged, all dissidents felt the pressure had played a significant part and the more foreign pressure the better.[13]

Over fifty victims examined by psychiatrists of the Moscow Working Commission between 1977 and 1981 and the files smuggled to the West by Vladimir Bukovsky in 1971 were the material which convinced most psychiatric associations that there was distinctly something wrong in the USSR.[12] :245

The Soviet authorities responded aggressively.[12] :45 Members of the group were being threatened, followed, subjected to house searches and interrogations.[12] :45 In the end, the members of the Commission were subjected to various terms and types of punishments: Alexander Podrabinek was sentenced to 5 years' internal exile, Irina Grivnina to 5 years' internal exile, Vyacheslav Bakhmin to 3 years in a labor camp, Dr Leonard Ternovsky to 3 years' labor camp, Dr Anatoly Koryagin to 8 years' imprisonment and labor camp and 4 years' internal exile, Dr Alexander Voloshanovich was sent to voluntary exile.[9] :153

In the autumn of 1978, the British Royal College of Psychiatrists carried a resolution in which it reiterated its concern over the abuse of psychiatry for the suppression of dissent in the USSR and applauded the Soviet citizens, who had taken an open stance against such abuse, by expressing its admiration and support especially for Semyon Gluzman, Alexander Podrabinek, Alexander Voloshanovich, and Vladimir Moskalkov.[14]

By the end of 1981, only Elena Bonner, Sofia Kalistratova and Naum Meiman were free, as a result of the unremitting campaign of persecution. The Moscow Helsinki Group was forced to cease operation.

The dissolution of the Moscow Helsinki Group was officially announced by Elena Bonner on 8 September 1982.[15] :35

## Rebirth of the group

However, in 1989, in the atmosphere of *glasnost*, it was re-established. A group of nine human rights activists, led by Larisa Bogoraz, the widow of Anatoly Marchenko, formally restarted the group on July 28, 1989. Included among the re-founders were Yuri Orlov and Lyudmila Alexeyeva, both part of the original group. Other prominent members are Larisa Bogoraz, Sergey Kovalev, Viatcheslav Bakhmin, Lev Timofeev, Henry Reznick, Lev Ponomarev, Gleb Yakunin, and Aleksei Simonov.

## See also

- Sofia Kalistratova
- International Helsinki Federation for Human Rights
- Helsinki Watch/Human Rights Watch

## References

[1]  "The Moscow Helsinki Group 30th Anniversary: From the Secret Files" (a selection of translated KGB/CPSU documents discussing MHG) http://www.gwu.edu/~nsarchiv/NSAEBB/NSAEBB191/index.htm

[2]  http://www.mhg.ru/history/14DB1C9

[3]  http://www.angelfire.com/sc3/soviet_jews_exodus/SovietNews_s/Gilbert_Secrets.shtml

[4]  http://www.eleven.co.il/article/15420

[5]  van Voren, Robert (2010). *Cold War in Psychiatry: Human Factors, Secret Actors* (http://books.google.com/books?id=Ru3-kQAACAAJ). Amsterdam—New York: Rodopi. ISBN 9042030461. .

[6]  Burns, John (26 July 1981). "Moscow silencing psychiatry critics" (http://www.nytimes.com/1981/07/26/world/ moscow-silencing-psychiatry-critics.html?pagewanted=print). *The New York Times*. . Retrieved 22 January 2011.

[7]  "The spread of Soviet suppression" (http://books.google.com/books?id=R0YnVUkTifgC&printsec=frontcover#PPA493,M1). *New Scientist* **78** (1104): 493. 25 May 1978. . Retrieved 22 January 2011.

[8]  Brintlinger, Angela; Vinitsky, Ilya (2007). *Madness and the mad in Russian culture* (http://books.google.com/ books?id=ED3U_XVLwHwC&printsec=frontcover#PPA92,M1). University of Toronto Press. ISBN 0802091407. .

[9]  *Medicine betrayed: the participation of doctors in human rights abuses* (http://books.google.com/books?id=bMTu_oIfVsIC& printsec=frontcover#PPA153,M1). Zed Books. 1992. pp. 153. ISBN 1856491048. .

[10]  van Voren, Robert; Bloch, Sidney (1989). *Soviet psychiatric abuse in the Gorbachev era* (http://books.google.com/ books?id=Q2xFAAAAYAAJ). International Association on the Political Use of Psychiatry. pp. 26. ISBN 9072657012. .

[11]  "Dr Alexander Voloshanovich: A Critic of the Political Misuse of Psychiatry in the USSR" (http://pb.rcpsych.org/cgi/reprint/4/5/70. pdf). *Psychiatric Bulletin* **4** (5): 70–71. 1980. doi:10.1192/pb.4.5.70. . Retrieved 20 January 2011.

[12]  van Voren, Robert (2009). *On Dissidents and Madness: From the Soviet Union of Leonid Brezhnev to the "Soviet Union" of Vladimir Putin* (http://books.google.ru/books?id=tyDIKu8XsgcC&printsec=frontcover#PPA61,M1). Amsterdam—New York: Rodopi. ISBN 9789042025851. .

[13]  Moran, Mark (19 November 2010). "Former Soviet Dissidents Believed APA Pressure Forced Change" (http://pn.psychiatryonline.org/ content/45/22/11.1.full). *Psychiatric News* **45** (22): 11. . Retrieved 20 January 2011.

[14]  "Autumn Quarterly Meeting 1978" (http://pb.rcpsych.org/cgi/reprint/3/1/5.pdf). *Psychiatric Bulletin* **3** (1): 5–7. 1979. doi:10.1192/pb.3.1.5. . Retrieved 23 January 2011.

[15]  Nuti, Leopoldo (2009). *The crisis of détente in Europe: from Helsinki to Gorbachev, 1975-1985* (http://books.google.com/ books?id=T3k9ednYVbwC&printsec=frontcover#PPA35,M1). Taylor & Francis. pp. 35. ISBN 0415460514. .

## Leaders

- Yuri Orlov (1976–1982)
- Larisa Bogoraz (1989–1993)
- Kronid Arkadyevich Lyubarsky (1993–1996)
- Lyudmila Michailovna Alekseyeva (since 1996)

## External links

- Moscow Helsinki Group (http://www.mhg.ru/english)
- Moscow Helsinki Group (http://www.civilsoc.org/nisorgs/russwest/moscow/helsinki.htm)
- Moscow Helsinki Group (MHG) (http://www.prison.org/english/ngomos.htm)

## Sources

- About the Netherlands Helsinki Committee (http://www.nhc.nl/about.php)
- *Документы Московской Хельсинкской Группы 1976—1982* (http://www.mhg.ru/files/knigi/docMHG.pdf)
- Alexeyeva, Ludmilla (1987). *Soviet dissent: contemporary movements for national, religious, and human rights* (http://books.google.com/books?id=5foGHAAACAAJ). Wesleyan University Press. ISBN 0819561762.
- Алексеева, Людмила (1992). *История инакомыслия в СССР: новейший период* (http://books.google.com/books?id=ucPXcAAACAAJ). Вильнюс—Москва: Весть. ISBN 5899422503. (The Russian text of the book in full is available online on the Memorial website by click (http://www.memo.ru/history/diss/books/ALEXEEWA/index.htm))
- *Власть и диссиденты: Из документов КГБ и ЦК КПСС* (http://www.mhg.ru/files/knigi/vendd.pdf). Москва: Московская Хельсинкская группа. 2006. ISBN 5984400340. (The Russian text of the book in full is available online on the MHG website by click (http://www.mhg.ru/files/knigi/vendd.pdf))

# Nikolai Chebotaryov

<table>
<tr><th colspan="2" align="center">Nikolai Chebotaryov</th></tr>
<tr><td colspan="2" align="center">Nikolai Chebotaryov (to the left) with pupils</td></tr>
<tr><td>Born</td><td>15 June 1894Kamianets-Podilskyi, Russian Empire (modern-day Ukraine)</td></tr>
<tr><td>Died</td><td>2 July 1947 (aged 53)Moscow, USSR</td></tr>
<tr><td>Nationality</td><td>Russian</td></tr>
<tr><td>Fields</td><td>Mathematics</td></tr>
<tr><td>Institutions</td><td>Kazan State University</td></tr>
<tr><td>Alma mater</td><td>Kiev State University</td></tr>
<tr><td>Doctoral advisor</td><td>Dmitry Grave</td></tr>
<tr><td>Doctoral students</td><td>Petr Kontorovich<br>Mark Krein<br>Naum Meiman<br>Vladimir Morozov</td></tr>
<tr><td>Known for</td><td>Chebotarev's density theorem</td></tr>
</table>

**Nikolai Grigorievich Chebotaryov** (often spelled *Chebotarov* or *Chebotarev*) (Russian: Никола́й Григо́рьевич Чеботарёв, Ukrainian: Микола Григорович Чоботарьов) (15 June [O.S. 3 June] 1894 − 2 July 1947) was a noted Russian and Soviet mathematician.[1] He is best known for the Chebotaryov density theorem.[2]

He was a student of Dmitry Grave, a famous Ukrainian and Russian mathematician.[3] Chebotaryov worked on the algebra of polynomials, in particular examining the distribution of the zeros. He also studied Galois theory and wrote an influential textbook on the subject titled *Basic Galois Theory*. He is additionally known for his work with his student Anatoly Dorodnov on the quadrature of the lune.[4]

On May 14, 2010, a memorial plaque for Nikolai Chebotaryov was unveiled on the main administration building of I.I. Mechnikov Odessa National University.

## References


[1]  O'Connor, John J.; Robertson, Edmund F., "Nikolai Grigorievich Chebotaryov" (http://www-history.mcs.st-andrews.ac.uk/Biographies/Chebotaryov.html), *MacTutor History of Mathematics archive*, University of St Andrews, ..

[2]  Lenstra, H. W.; Stevenhagen, P. (1996), "Chebotarëv and his density theorem" (http://websites.math.leidenuniv.nl/algebra/chebotarev.pdf), *Mathematical Intelligencer* **18** (2): 26–37, doi:10.1007/BF03027290, .

[3]  Nikolay Grigorievich Chebotarev (http://genealogy.math.ndsu.nodak.edu/id.php?id=73922) at the Mathematics Genealogy Project.

[4]  Postnikov, M. M. (2000), "The problem of squarable lunes", *American Mathematical Monthly* **107** (7): 645–651, JSTOR 2589121. Translated from Postnikov's 1963 Russian book on Galois theory.

# Doktor nauk

**Doktor nauk** (Russian: доктор наук, literally translated as "Doctor of Sciences") is a higher doctoral degree, the second and the highest post-graduate academic degree in the Soviet Union, Russia and in many post-Soviet states. Sometimes referred to as Dr. Hab. The prerequisite is the first degree, Kandidat nauk (Russian: кандидат наук, lit. "candidate of sciences") which is informally regarded equivalent to Ph.D. degree [1] . Both *Doktor* and *Kandidat* must meet nation-wide standards enforced by the Russian federal government.

Doktor Nauk is conferred by the federal government agency, the Higher Attestation Commission (Russian: Высшая аттестационная комиссия, abbreviated Russian: ВАК, **VAK**), on the solicitation by the Specialized Dissertation Committee where the doctoral candidate had defended his/her research work. Such committees are created in academic institutions with established research record and are accredited by VAK.[2] The total number of committee members is typically about 20, all holding the Doktor Nauk degree. The area of research specialization of at least five committee members must match the profile of the materials submitted by the doctoral candidate for the consideration. The candidate must conduct independent research. Therefore, no academic supervisor is required; moreover, typically the candidate is an established scholar him/herself, supervising a few Ph.D. students while working towards his/her Doktor Nauk. However, it is normal practice when an experienced consultant is appointed to help the scholar with identifying the research problem and finding the approach to solving it; yet this is not technically regarded a supervision.

The procedures of conferring of both *Kandidat* and *Doktor* academic degrees are more formal and different from conferring a Ph.D. degree in Western universities. In particular, for the *Doktor*, the academic institution, where the scholar is affiliated as a doctoral candidate, must conduct a preliminary review of the research results and personal contribution made by the candidate and, depending on findings, elect whether to render formal support or not. By definition, this highly prestigious degree can be conferred only for a significant contribution to science and/or technology based on a public defense of a thesis, monograph, or (in rare cases) of a set of outstanding publications in referreed journals. The defense must be held at the session of a Specialized Dissertation Committee accredited by VAK. Prior to the defense, three referees holding Doktor Nauk degrees themselves (the so-called "official opponents"), must submit their written motivated assessments of the thesis. One more similar assessment is to be provided by some university or academic institution, working in the same field of science or technology, and in addition several other reviewers must mail their conclusions made based on a thesis summary (usually a 32 page brochure).

In the former USSR, this degree is considered sufficient credential for tenured full professorship at any institution of higher learning. Unless an academic holds a Doktor Nauk, she/he can make it to a full professor only through 15 years or more of outstanding teaching service on the university level. At least one published and widely accepted textbook and the degree of Kandidat Nauk are required in the latter case, anyway. A Doktor Nauk degree holder can become a tenured full professor after just one year of teaching experience in a non-tenured faculty position.

The Doktor Nauk thus has no academic equivalent in North America, as it is a post-doctoral degree. The German Habilitation and, to some extent, the French "habilitation à diriger des recherches" (HDR) are comparable to it, as are the British senior doctorates (e.g. Doctor of Science), although the latter hold little contemporary relevance. On the average, only 10 per cent of Kandidats eventually earn a Doktor degree. Although some exceptionally talented researchers in Mathematics do earn Doktor Nauk in their late 20s, the average age of the scholars reaching Doktor in most disciplines is about 50; this implicitly indicates the amount of contribution that must be made.

## See also

- Education in Russia
- Habilitation
- Academic degree

## References

[1]  Educational Records Evaluation Service (http://www.eres.com)
[2]  NIC ARaM of the Ministry of Education of the Russian Federation (http://www.russianenic.ru/english/rus/index.html#8)

# University of Kharkiv

<table>
<tr><td colspan="2" align="center">Kharkiv National University</td></tr>
<tr><td colspan="2" align="center">Харківський національний університет імені В. Н. Каразіна</td></tr>
<tr><td colspan="2" align="center">Latin: Universitas Charkoviensis</td></tr>
<tr><td>Motto</td><td>Cognoscere, Docere, Erudire</td></tr>
<tr><td>Motto in English</td><td>To learn, to educate, to enlighten</td></tr>
<tr><td>Established</td><td>1804</td></tr>
<tr><td>Type</td><td>Public university</td></tr>
<tr><td>President</td><td>Vil S. Bakirov</td></tr>
<tr><td>Academic staff</td><td>1,300</td></tr>
<tr><td>Students</td><td>15,000</td></tr>
<tr><td>Postgraduates</td><td>500</td></tr>
<tr><td>Location</td><td>Kharkiv, Ukraine</td></tr>
<tr><td>Colors</td><td>Blue and White</td></tr>
<tr><td>Affiliations</td><td>IAU, EUA</td></tr>
<tr><td>Website</td><td>univer.kharkov.ua [1]</td></tr>
</table>

The **University of Kharkiv** (Ukrainian: Харківський університет) or officially the **Karazin Kharkiv National University** (Ukrainian: Харківський національний університет імені В. Н. Каразіна) is one of the major universities in Ukraine, and earlier in the Russian Empire and Soviet Union. It was founded in 1804 through the efforts of Vasyl Karazin becoming the second oldest university in Ukraine after the University of Lviv.

## Ranking

In 2010, according to University Ranking by Academic Performance (URAP),[2] it is the best university in Ukraine and 1415th university in the world.

# Units

## Departments

- Department of Biology
- Department of Chemistry
- Department of Ecology
- Department of Economics
- Department of International Economic Relations and Tourist Industry
- Department of Foreign Languages
- Department of Medicine [3]
- Department of Geology and Geography
- Department of History
- Department of Law
- Department of Mathematics and Mechanical Engineering
- Department of Physics
- Department of Philology
- Department of Philosophy
- Department of Psychology
- Department of Radiophysics
- Department of Sociology

## In the frame of the Institute of High Technologies

- Department of Physics and Technology
- Department of Computer Science
- Department of Energy Physics

The University of Kharkiv main academic building

The northern academic building

# Notable alumni

- Andrei Kolegayev

Among the scholars associated with the university there are two Nobel prize winners:

- Élie Metchnikoff (Medicine, 1908)
- Lev Landau (Physics, 1962)

## References

[1] http://www.univer.kharkov.ua/en

[2] University Ranking by Academic Performance (2010). "Rank by Country. Ukraine" (http://www.urapcenter.org/2010/country. php?ccode=UA). . Retrieved 2011-02-26.

[3] http://md.univer.kharkov.ua/

## See also

- List of modern universities in Europe (1801–1945)

## External links

- Official Website (http://www.univer.kharkov.ua/en) (English) (Ukrainian)
- University of Kharkiv (http://www.encyclopediaofukraine.com/display. asp?AddButton=pages\K\H\KharkivUniversity.htm) at Encyclopedia of Ukraine
- University Library (http://www-library.univer.kharkov.ua/) (English) (Ukrainian) (Russian)

# Lev Landau

<table>
<tr><td colspan="2" align="center">Lev Landau</td></tr>
<tr><td colspan="2" align="center">Lev Davidovich Landau (1908-1968)</td></tr>
<tr><td>Born</td><td>January 22, 1908Baku, Russian Empire</td></tr>
<tr><td>Died</td><td>April 1, 1968 (aged 60)Moscow, Soviet Union</td></tr>
<tr><td>Residence</td><td>Soviet Union</td></tr>
<tr><td>Citizenship</td><td>Soviet Union</td></tr>
<tr><td>Fields</td><td>Theoretical Physics</td></tr>
<tr><td>Institutions</td><td>Baku State University<br>Kharkiv University<br>Kharkiv Polytechnical Institute<br>Institute for Physical Problems<br>MSU Faculty of Physics</td></tr>
<tr><td>Alma mater</td><td>Leningrad State University<br>Leningrad Physico-Technical Institute</td></tr>
<tr><td>Doctoral students</td><td>Alexei Alexeyevich Abrikosov<br>Isaak Markovich Khalatnikov</td></tr>
<tr><td>Other notable students</td><td>Evgeny Lifshitz</td></tr>
<tr><td>Known for</td><td>Superfluidity<br>Superconductivity<br>Course of Theoretical Physics</td></tr>
<tr><td>Notable awards</td><td>Stalin Prize (1946)<br>Nobel Prize in Physics (1962)</td></tr>
</table>

**Lev Davidovich Landau** (Russian language: Лёв Дави́дович Ланда́у; January 22 [O.S. January 9] 1908 – April 1, 1968) was a prominent Soviet physicist who made fundamental contributions to many areas of theoretical physics. His accomplishments include the independent co-discovery of the density matrix method in quantum mechanics (alongside John von Neumann), the quantum mechanical theory of diamagnetism, the theory of superfluidity, the theory of second-order phase transitions, the Ginzburg–Landau theory of superconductivity, the theory of Fermi liquid, the explanation of Landau damping in plasma physics, the Landau pole in quantum electrodynamics, and the two-component theory of neutrinos. He received the 1962 Nobel Prize in Physics for his development of a mathematical theory of superfluidity that accounts for the properties of liquid helium II at a temperature below 2.17 K (−270.98 °C).

# Biography

### Early years

Landau was born on January 22, 1908 to a Jewish family[1] in Baku, in what was then the Russian Empire. Landau's father was an engineer with the local oil industry and his mother was a doctor. Recognized very early as a child prodigy in mathematics, Landau was quoted as saying in later life that he scarcely remembered a time when he was not familiar with calculus. Landau graduated at 13 from gymnasium. His parents considered him too young to attend university, so for a year he attended the Baku Economical Technicum. In 1922, at age 14, he matriculated at Baku State University, studying in two departments simultaneously: the department of Physics and Mathematics, and the department of Chemistry. Subsequently he ceased studying chemistry, but remained interested in the field throughout his life.

In 1924, he moved to the main centre of Soviet physics at the time: the Physics Department of Leningrad State University. In Leningrad, he first made the acquaintance of genuine theoretical physics and dedicated himself fully to its study, graduating in 1927. Landau subsequently enrolled for post-graduate study at the Leningrad Physico-Technical Institute, and at 21, received a doctorate. Landau got his first chance to travel abroad in 1929, on a Soviet government traveling fellowship supplemented by a Rockefeller Foundation fellowship.

After brief stays in Göttingen and Leipzig, he went to Copenhagen to work at Niels Bohr's Institute for Theoretical Physics. After the visit, Landau always considered himself a pupil of Niels Bohr and Landau's approach to physics was greatly influenced by Bohr. After his stay in Copenhagen, he visited Cambridge and Zürich before returning to the Soviet Union. Between 1932 and 1937 he headed the department of theoretical physics at the Kharkov Polytechnic Institute.

### Great Purge

During the Great Purge, Landau was investigated within the UPTI Affair in Kharkov, but he managed to leave for Moscow. Still, he was arrested on April 27, 1938 and held in an NKVD prison until his release on April 29, 1939, after his colleague Pyotr Kapitsa, an experimental low-temperature physicist, wrote a letter to Joseph Stalin, personally vouching for Landau's behavior. Gorelik, in a *Scientific American* article in 1997, described Landau's life and interactions with the Soviet intelligence agency during Stalin era and post-Stalin phase.

### Last years

On January 7, 1962, Landau's car collided with an oncoming truck. He was severely injured and spent two months in a coma. Although Landau recovered in many ways, his scientific creativity was destroyed, and he never returned fully to scientific work. His injuries prevented him from accepting the 1962 Nobel Prize for physics in person.[2]

In 1965 former students and coworkers of Landau founded the Landau Institute for Theoretical Physics, located in the town of Chernogolovka near Moscow, and headed for the following three decades by Isaak Markovich Khalatnikov.

## Death

Landau died on April 1, 1968, aged 60, from complications of the injuries from the car accident he was involved in 6 years earlier. He was buried at Novodevichy cemetery.[3] [4]

## The Landau School

Apart from his theoretical accomplishments, Landau was the principal founder of a great tradition of theoretical physics in Kharkov, Soviet Union (now Kharkiv, Ukraine), sometimes referred to as the "Landau school". He was the head of the Theoretical Division at the Institute for Physical Problems from 1937 until 1962 when, as a result of a car accident, he suffered injuries which stopped him from making further contributions to science.[5] His students included Lev Pitaevskii, Alexei Abrikosov, Arkady Levanyuk, Evgeny Lifshitz, Lev Gor'kov, Isaak Khalatnikov, Boris L. Ioffe, Roald Sagdeev and Isaak Pomeranchuk.

Landau developed a comprehensive exam called the "Theoretical Minimum" which students were expected to pass before admission to the school. The exam covered all aspects of theoretical physics, and between 1934 and 1961 only 43 candidates passed.

In Kharkov, he and his friend and former student, Evgeny Lifshitz, began writing the *Course of Theoretical Physics*, ten volumes that together span the whole of the subject and are still widely used as graduate-level physics texts.

## Legacy

Two celestial objects are named in his honor:

- the minor planet 2142 Landau.[6]
- the lunar crater Landau.

## Landau's List

Landau kept a list of names of physicists which he ranked on a logarithmic scale of productivity ranging from 0 to 5. The highest ranking, 0.5, was assigned to Albert Einstein. A rank of 1 was awarded to "historical giants" Isaac Newton, Eugene Wigner, and the founding fathers of quantum mechanics, Niels Bohr, Werner Heisenberg, Paul Dirac and Erwin Schrödinger. Landau ranked himself as a 2.5 but later promoted himself to a 2. David Mermin, writing about Landau, referred to the scale, and ranked himself in the fourth division, in the article *My Life with Landau: Homage of a 4.5 to a 2.*[7] [8]

Commemorative plaque in Baku.

## Works

### Landau and Lifshitz *Course of Theoretical Physics*

- L.D. Landau, E.M. Lifshitz (1976). *Mechanics*. **Vol. 1** (3rd ed.). Butterworth-Heinemann. ISBN 978-0-750-62896-9.
- L.D. Landau, E.M. Lifshitz (1975). *The Classical Theory of Fields*. **Vol. 2** (4th ed.). Butterworth-Heinemann. ISBN 978-0-750-62768-9.
- L.D. Landau, E.M. Lifshitz (1977). *Quantum Mechanics: Non-Relativistic Theory*. **Vol. 3** (3rd ed.). Pergamon Press. ISBN 978-0-080-20940-1.
- V.B. Berestetskii, E.M. Lifshitz, L.P. Pitaevskii (1982). *Quantum Electrodynamics*. **Vol. 4** (2nd ed.). Butterworth-Heinemann. ISBN 978-0750633710.

- L.D. Landau, E.M. Lifshitz (1980). *Statistical Physics, Part 1*. **Vol. 5** (3rd ed.). Butterworth-Heinemann. ISBN 978-0-750-63372-7.
- L.D. Landau, E.M. Lifshitz (1987). *Fluid Mechanics*. **Vol. 6** (2nd ed.). Butterworth-Heinemann. ISBN 978-0-080-33933-7.
- L.D. Landau, E.M. Lifshitz (1986). *Theory of Elasticity*. **Vol. 7** (3rd ed.). Butterworth-Heinemann. ISBN 978-0-750-62633-0.
- L.D. Landau, E.M. Lifshitz, L.P. Pitaevskii (1984). *Electrodynamics of Continuous Media*. **Vol. 8** (1rst ed.). Butterworth-Heinemann. ISBN 978-0-750-62634-7.
- L.P. Pitaevskii, E.M. Lifshitz (1980). *Statistical Physics, Part 2*. **Vol. 9** (1rst ed.). Butterworth-Heinemann. ISBN 978-0-750-62636-1.
- L.P. Pitaevskii, E.M. Lifshitz (1981). *Physical Kinetics*. **Vol. 10** (1rst ed.). Pergamon Press. ISBN 978-0-750-62635-4.

## Other books

- L.D. Landau, A.J. Akhiezer, E.M. Lifshitz (1967). *General Physics, Mechanics and Molecular Physics*. Pergamon Press. ISBN 978-0-080-09106-8.

## Books about Landau

- Dorozynski, Alexander (1965). *The Man They Wouldn't Let Die*. Secker and Warburg. ASIN B0006DC8BA. (After Landau's 1962 car accident, the physics community around him rallied to attempt to save his life. They managed to prolong his life until 1968.)
- Janouch, Frantisek (1979). *Lev D. Landau: His life and work*. CERN. ASIN B0007AUCL0.
- Khalatnikov, I.M., ed (1989). *Landau. The physicist and the man. Recollections of L.D. Landau*. Sykes, J.B. (trans.). Pergamon Press. ISBN 0-08-036383-0.
- Kojevnikov, Alexei B. (2004). *Stalin's Great Science: The Times and Adventures of Soviet Physicists*. History of Modern Physical Sciences. Imperial College Press. ISBN ISBN 1-86094-420-5.
- Landau-Drobantseva, Kora (1999). *Professor Landau: How We Lived* [9]. AST. ISBN 5-8159-0019-2. (Russian)

## In popular culture

- Russian television film *My Husband - the Genius* (unofficial translation of the Russian title *Мой муж — гений*) released in 2008 tells biography of Landau (played by Daniil Spivakovsky), mostly relying on his private life. It was generally panned by critics. People who had personally met Landau, including famous Russian scientist Vitaly Ginzburg, said that the film was not only terrible but also false in historical facts.
- Another film about Landau, *Dau*, is in the filming process now and was originally planned to be released in 2010. It is directed by Ilya Khrzhanovsky with non-professional actor Theodor Kurentzis (an orchestra conductor) as Landau.

## See also

- Landau–Hopf theory of turbulence
- Landau–Lifshitz–Gilbert equation
- Landau–Lifshitz model
- Landau (crater)
- Landau theory of second order phase transitions
- Ginzburg–Landau theory of superconductivity
- Landau quantization, Landau levels
- Landau damping
- List of Jewish Nobel laureates

## References

[1]  *The Jews in the twentieth century: an illustrated history*, by Martin Gilbert, Schocken Books, 2001, page 284

[2]  Nobel Presentation speech by Professor I. Waller, member of the Swedish Academy of Sciences (http://nobelprize.org/nobel_prizes/ physics/laureates/1962/press.html)

[3]  www.findagrave.com (http://www.findagrave.com/cgi-bin/fg.cgi?page=gr&GRid=8050669)

[4]  novodevichye.com (http://novodevichye.com/landau/)

[5]  Alexander Dorozynski (1965). *The Man They Wouldn't Let Die*.

[6]  Lutz D. Schmadel (2003). *Dictionary of Minor Planet Names* (http://books.google.com/books?q=2141+Simferopol+1970) (5th ed.). Springer Verlag. pp. 174. ISBN 3-540-00238-3. .

[7]  Tony Hey (1997). *Einstein's Mirror*. Cambridge University Press. pp. 1. ISBN 0-521-43532-3.

[8]  Asoke Mitra; Ramlo, Susan; Dharamsi, Amin; Mitra, Asoke; Dolan, Richard; Smolin, Lee (2006). "New Einsteins Need Positive Environment, Independent Spirit". *Physics Today* **59** (11): 10. Bibcode 2006PhT....59k..10H. doi:10.1063/1.2435630.

[9]  http://lib.ru/MEMUARY/LANDAU/landau.txt

- Karl Hufbauer, "Landau's youthful sallies into stellar theory: Their origins, claims, and receptions," Historical Studies in the Physical and Biological Sciences, 37 (2007), 337–354.

## Further reading

- O'Connor, John J.; Robertson, Edmund F., "Lev Landau" (http://www-history.mcs.st-andrews.ac.uk/ Biographies/Landau_Lev.html), *MacTutor History of Mathematics archive*, University of St Andrews.

- Lev Davidovich Landau (http://www.nobel-winners.com/Physics/lev_davidovich_landau.html)

- Landau's Theoretical Minimum, Landau's Seminar, ITEP in the Beginning of the 1950's (http://arxiv.org/abs/ hep-ph/0204295v1) by Boris L. Ioffe, Concluding talk at the workshop *QCD at the Threshold of the Fourth Decade/Ioeffest*.

- EJTP Landau Issue 2008. (http://www.ejtp.info/landau.html)

- Ammar Sakaji and Ignazio Licata (eds), Lev Davidovich Landau and his Impact on Contemporary Theoretical Physics (https://www.novapublishers.com/catalog/product_info.php?products_id=9382), Nova Science Publishers, New York, 2009, ISBN 978-1-60692-908-7.

- Gennady Gorelik, The top-secret life of Lev Landau, *Scientific American*, Aug. 1997, vol.277(2), 53–57.

# Article Sources and Contributors

**Naum Meiman** *Source*: http://en.wikipedia.org/w/index.php?title=Naum_Meiman *Contributors*: Alex Bakharev, INeverCry, Ironholds, Lkitrossky, Marjaliisa, Mhym, Michael Hardy, Omnipaedista, Sodin

**Soviet dissidents** *Source*: http://en.wikipedia.org/w/index.php?title=Soviet_dissidents *Contributors*: Altenmann, Arf!, BACbKA, Black Falcon, CieloEstrellado, Domitori, FeanorStar7, Hmains, Hodja Nasreddin, Hugh16, Ilyaroz, Ldingley, MILH, Meishern, Michael Hardy, Muscovite99, Olegwiki, Ostap R, P.o.l.o., Philip Cross, Psychiatrick, Qmwne235, Ramir, Rglovejoy, Samian, The Nut, VVPushkin, 16 anonymous edits

**Baranovichi** *Source*: http://en.wikipedia.org/w/index.php?title=Baranovichi *Contributors*: AL PAÇIO, Ace Class Shadow, Alex756, AntBel, Atitarev, Bacian, Balcer, BoH, CommonsDelinker, Czalex, DDima, Darwinek, DerBorg, Dr. Blofeld, Dr. Dan, Emax, EugeneZelenko, Future Perfect at Sunrise, Ghirlandajo, Halibutt, HennessyC, Igor Mostitsky, Irpen, KPbIC, Kuban kazak, Kwamikagami, LHOON, Languagehat, Ld, Leutha, MER-C, Maksim L., Marek69, MartinDK, Metsavend, Monegasque, Monkbel, Narking, Nk, Omar35880, Picapica, Pleckaitis, R9tgokunks, RM241, Raviaka Ruslan, Rocastelo, Romanskolduns, The wub, Timrollpickering, Truthseeker 85.5, Tymek, Unomano, Unyoyega, Vadim Akopyan, Voyevoda, Wv participant, Ximar, Zscout370, 36 anonymous edits

**Mathematician** *Source*: http://en.wikipedia.org/w/index.php?title=Mathematician *Contributors*: 0, 1exec1, 203.26.98.xxx, 28421u2232nfenfcenc, 63.192.137.xxx, A little insignificant, A4, Aarsalankhalid, Addshore, Ahoerstemeier, Ajraddatz, Akjar13, Alansohn, Alexandria, Almit39, Alteration10, Ancheta Wis, Andmatt, Andre Engels, Animum, Anonymi, Anonymous Dissident, Anonymous the Editor, Antandrus, Ante Aikio, Anthony Appleyard, Anton Gutsunaev, Apptas, Aqwis, Architectual, Arjun01, Avoided, AxelBoldt, BD2412, Babalapaloop, Banej, BarretB, Bassbonerocks, Bbggae, Belovedfreak, Bevo, Bfigura's puppy, Bhawani Gautam, BlaiseFEgan, Bliduta, Blinking Spirit, Blotwell, Bobo192, Bongwarrior, Bootyy, Booyabazooka, Brandmeister (old), Brianjd, Brion VIBBER, ButOnMethItIs, Buzybeez, C S, C.Fred, CBDunkerson, CBM, CRGreathouse, Cafeturque, Calabe1992, Calypso, CambridgeBayWeather, Camilo Sanchez, Canderson7, Canuckistan Rocks, Capitalist, Catmoongirl, Cenarium, Chapeaubien, Chas zzz brown, ChicXulub, Chrisbehk, Ciphers, Cj67, Closedmouth, Conditionhumaine, Conversion script, Coolman2k8, Correogsk, Cpl Syx, CptCutLess, Cronholm144, D.M. from Ukraine, DARTH SIDIOUS 2, DSachan, Damirgraffiti, Daniel Case, Darnell24, Darolew, David Eppstein, David.Monniaux, DavidLevinson, Db099221, Dbenbenn, Dbroadwell, Dcclark, DealPete, Demmy100, DerHexer, Derickjay, Deus Ex, Deville, Dicklyon, DigitalCharacter, Diligent Terrier, Dlohcierekim, Docu, Dominus, DonaNobisPacem, Donarreiskoffer, Doulos Christos, DoversCHAMP, Drmies, Drosdaf, Dsgdfshfdshdsfh, Eastlaw, Eclecticology, Edcolins, Edison, Eeekster, Elockid, Epbr123, Epsilon60198, Eric119, Etale, Eternityflow, Everyking, Excirial, FF2010, FJPB, Falcon8765, Favonian, Fephisto, Foljiny, Fredrik, Frentz, Fyyer, Fæ, GTBacchus, Galoubet, Gap9551, Gauss, Geneffects, Georg Muntingh, Gguidelas, Giftlite, Gilliam, Gioto, Glane23, Gogo Dodo, Goodnightmush, Googl, Gpvos, Grafen, Graham87, Grammar conquistador, Greg Kuperberg, Grim23, Grimmcar, Grinder0-0, Gurchzilla, Haham hanuka, HamburgerRadio, Hdt83, HedgeHog, Hemmingsen, Hennessey, Patrick, Henrik, Heracles31, Hu12, I dream of horses, ICanAlwaysChangeThisLater, Iain99, Ianw17, Ibbn, Igoldste, IndieAddiction, Insanity Incarnate, Iothiania, Iriseyes, Irˀtiˀnal, Ixfd64, J'88, J.delanoy, J271828, JDspeeder1, JForget, Jackfork, Jackol, Jacobolus, Jagged 85, Jahiegel, Jaimeastorga2000, Jallotta, JamesMLane, Jamesofur, JamieJones, Jankur, Janlo, Jcm, Jeff G., Jeffrey Mall, Jeffrey O. Gustafson, Jeffroyracer, Jeronimo, Jfhgja;, JillandJack, Jim.belk, JimWae, Jitse Niesen, Joe Fabeetz, Joeldudesx2, Joerite, John254, Jonas Mur, Joseph Solis in Australia, Joshua Issac, JoshuaZ, Jstnmgr, Juan Marquez, Juliancolton, Jusdafax, Justin W Smith, Jvalure, Karada, KarlHallowell, Katherine, Kazikame, Kehrbykid, Kinaro, Kindyin, King of Hearts, Kingbusch, Kkm010, Koavf, Konstable, Koyaanis Qatsi, Kubigula, Kudret abi, Kungfuadam, Kungming2, Kuru, LA2, LJosil, Laker9903, Lambiam, Leon math, Leuko, Lihaas, Little guru, LittleDan, Logicalasif, Lradrama, Lupin, Lupo, Lyleq, LéonTheCleaner, MKoltnow, Madhero88, Maelnuneb, Maidenindigo, Mannafredo, Marcika, Marek69, Master of Puppets, Masterpiece2000, Materialscientist, MathKnight, Mathmo, Matthead, Maurice Carbonaro, MegaPedant, Menareretards, Meni Rosenfeld, Meno25, Mets501, Mhym, Michael Hardy, Michael Slone, Micione, Mikez, Minimac's Clone, Miquonranger03, Mir76, Misza13, Mmnaw, Moderate2008, Morios, Mossgiantkiller, Moverton, Mrdthree, Mtking, Myasuda, Myopic Bookworm, Mysteronald, N5iln, Nburden, Nbvjou, Nepenthes, Netoholic, Nev1, Newkai, Nick 1234567, NikolaiLobachevsky, Nixdorf, Nuadh, Nwwaew, Ocaasi, Ocolon, Oleg Alexandrov, Olivier, Omicronpersei8, Orange Suede Sofa, Out of Phase User, Oğuz Ergin, PMajer, Palakisbest, Pan Dan, Patrick, Patrick Berry, Paul August, Paw7151, Pdcook, Persian Poet Gal, Pete.Hurd, Peterlin, Petrb, Pgecaj, PhS, Phantomsteve, Philip Trueman, Piano non troppo, Pichu826, Pinethicket, Pleasantville, Point-set topologist, Poor Yorick, Poorleno, Pred, Pricebank, Prolog, Proofreader77, Prunesqualer, Psychonaut, Pupster21, Quentar, Quuxplusone, R. fiend, RDBury, RMcGuigan, Raepetersen, Reach Out to the Truth, Reddi, Redfarmer, Redman775, Reedy, Requestion, Rhythm, Rich Farmbrough, Richard L. Peterson, RjLesch, RobLa, Robert W. Wright, RobertG, Robinh, Ronhjones, Rotem Dan, Rrokkedd, Rsrikanth05, Ryulong, SCZenz, SU Linguist, Sam Korn, Samuel, Samuel Bach, Sandman, Sarenne, Satori Son, Schneelocke, SchoolPlayAround3, SchuminWeb, Scohoust, Sdornan, Seaphoto, Semperf, Sepastaj, Shadowjams, Shanes, Shimmin, Shoeofdeath, Shogartu, Sibian, SimonP, Sinadoom, Sj, Smalljim, Sp33dyphil, SpuriousQ, Srich32977, Srisou, StaticGull, Stephenb, Strojo4, Suklaa, Sushiflinger, Synchronism, TakuyaMurata, Tangent747, Tawker, Taxman, Tbackstr, Teeth123123, Terrek, Tesseran, Tetraedycal, The Anome, The Fish, The Thing That Should Not Be, Themanofthehouserulzzz, Therepel, Thingg, Tide rolls, TigerShark, Tim1357, Timwi, To Fight a Vandal, Tobias Bergemann, Toddst1, Tom Morris, Tomgally, Tomgreeny, Tommy2010, Tony pal, TonyBallioni, TooMuchMath, Topology Expert, Tosayit, Trang Oul, Typuifre, Ukexpat, Ulric1313, UltimateKingSatan, Uncle Dick, Uncool95, UnitedStatesian, Useight, Vanished User 1004, Versus22, Viriditas, Vishnava, Vornoff, Vrenator, Vsmith, Vytautas, Wafulz, Waltervulej, Wayward, Weerasad, Whaa?, Why Not A Duck, WikHead, WikipedianMarlith, Wikipelli, Wilanthule, William Avery, Winchelsea, Wine Guy, Wlod, Wolfrock, Wtmitchell, Wuffyz, Wwannsda, Wysprgr2005, XJamRastafire, Xp54321, Yamamoto Ichiro, Yankeetown12, Ybbor, Yfkmafiaboss, Z3dman, ZX81, Zmoney918, Zomno, ZooFari, Zsinj, Zundark, Zvika, Zzyxzaa26, Александър, 1085 anonymous edits

**Complex analysis** *Source*: http://en.wikipedia.org/w/index.php?title=Complex_analysis *Contributors*: 63.162.153.xxx, 9258fahsflkh917fas, ABCD, Aarishshaheen, Abdull, Adam majewski, Ahoerstemeier, Aitias, Ali Obeid, AlsatianRain, AndreasB, AnnaFrance, Anonymous Dissident, AxelBoldt, BigJohnHenry, Blehfu, Blindsuperhero, Brad7777, Brian Tvedt, CJauff, Charles Matthews, Conversion script, Crayolacrime, Cwkmail, DARTH SIDIOUS 2, Daniel Brockman, DavidCBryant, Didactik, Digby Tantrum, DivideByZero14, Dratman, Dysprosia, Edokter, El C, Excirial, Favonian, Fiona CS, Fredrik, Ft1, Gamewizard71, Gareth Wyn, Giftlite, Haiviet, Hesperian, Ineffable3000, Isnow, Jim.belk, Jitse Niesen, Jkimath, Jonatan Swift, Jowa fan, Jusdafax, Jóna Þórunn, Karl-Henner, Laurent MAYER, LokiClock, MCrawford, Mana Excalibur, Maurice Carbonaro, Maxim, Mboverload, Michael Hardy, Michael Slone, Msh210, Nauticashades, Neilc, Obradovic Goran, Oleg Alexandrov, Paul August, Paul D. Anderson, Penubag, Piet Delport, Point-set topologist, Raven4x4x, Rich Farmbrough, RobHar, Robert Illes, Rocchini, Rock69, Rockyrackoon, Ruud Koot, Schneelocke, Sjoerd visscher, Sverdrup, TakuyaMurata, Taxman, Tayoun7, Thatoneguy, Tony Fox, Vice regent, Vvelup, דוד ש׳, 127 anonymous edits

**Partial differential equation** *Source*: http://en.wikipedia.org/w/index.php?title=Partial_differential_equation *Contributors*: Affluent Rider, Ahoerstemeier, Aliotra, Alpha Quadrant (alt), Andrei Polyanin, AndrewHowse, Arnero, ArnoldReinhold, Arthena, AxelBoldt, Belovedfreak, Bemoeial, Ben pcc, BenFrantzDale, Bender235, Bertik, Bjorn.sjodin, Borgx, Brian Tvedt, CYD, Cbm, Charles Matthews, Chbarts, Chris in denmark, ChristophE, Cj67, Ckatz, Crowsnest, Crust, CyrilB, D.328, DStoykov, David Crawshaw, Dharma6662000, Dicklyon, Dirkbb, Djordjes, DominiqueNC, Donludwig, DrHok, Dysprosia, Egriffin, Eienmaru, Eigenlambda, El C, EmmetCaulfield, Epbr123, Erxnmedia, Evankeane, Filemon, Fintor, Foober, Frosted14, Fuse809, Gaj0129, Gerasime, Germandemat, Giese, Giftlite, GraemeL, Gseryakov, Gurch, HappySophie, Hongooi, Hut 8.5, Isnow, Iwfyita, Ixfd64, JNW, JaGa, Jitse Niesen, Jmath666, Jon Cates, JonMcLoone, Jonathanstray, Jss214322, Jyril, Kbolino, Kwiki, L-H, Linas, MFH, Magister Mathematicae, Mandolinface, Manticore, MathMartin, Mathsfreak, Maurice Carbonaro, Mazi, Mhaitham.shammaa, Mhym, Michael Devore, Michael Hardy, Moink, Mpatel, Msh210, MuthuKutty, NSiDms, Nbarth, Nneonneo, Ojcit, Oleg Alexandrov, Oliver Pereira, OrgasGirl, Oscarjquintana, PL290, Pacaro, Patrick, Paul August, Paul Matthews, PeR, PhotoBox, Pokespa, Pranagailu1436, Prime Entelechy, Pt, Quibik, R'n'B, R.e.b., Rausch, RayAYang, Richard77, Rjwilmsi, Rnt20, Roadrunner, Robinh, Roesser, Rpchase, Salih, Sbarnard, Siegmaralber, SobakaKachalova, Spartan-James, Srleffler, Stevenj, Stizz, Super Cleverly, Sławomir Biały, THEN WHO WAS PHONE?, Tarquin, Tbsmith, The Anome, The Transhumanist, Thenub314, Tiddly Tom, Timwi, Topbanana, Tosha, Ub3rm4th, Unigfjkl, User A1, Waltpohl, Wavelength, Wavesmikey, Winston365, Wolfrock, Wsulli74, Wtt, Yaje, Yhkhoo, Zhou Yu, Zzuuzz, 303 anonymous edits

**Moscow Helsinki Group** *Source*: http://en.wikipedia.org/w/index.php?title=Moscow_Helsinki_Group *Contributors*: Abune, Alex Bakharev, AvicAWB, CieloEstrellado, Domitori, Dzied Bulbash, Ghirlandajo, Gritzko, Hodja Nasreddin, Jnc, Kam Solusar, Larvatus, Pax:Vobiscum, Philip Cross, Piast93, Psychiatrick, Redfox24, Renata3, Russavia, Tec15, Yuriybrisk, 7 anonymous edits

**Nikolai Chebotaryov** *Source*: http://en.wikipedia.org/w/index.php?title=Nikolai_Chebotaryov *Contributors*: Ahonc, Amalas, Bender235, Catgirl, Charles Matthews, David Eppstein, Eaefremov, Everyking, God of Sins, Ian Pitchford, Jsqqq777, Ketiltrout, Lkitrossky, M.V.E.i., Mhym, Omnipaedista, Ru.spider, Sergey Marchenko, Waacstats, Wmahan, Zaslav, 8 anonymous edits

**Doktor nauk** *Source*: http://en.wikipedia.org/w/index.php?title=Doktor_nauk *Contributors*: Aleksa Lukic, Altenmann, Brokenwit, Butseriouslyfolks, CanadianLinuxUser, Cassowary, Cripipper, DGG, Dlazerka, Fran Rogers, Infovarius, Jakkm77, Kaihsu, Loren.wilton, Maxal, Mzabduk, Nneonneo, Nv8200p, Olaffpomona, Omnipaedista, Rjwilmsi, SilentHussar, Tr00rle, Unionhawk, Venediktov, 21 anonymous edits

**University of Kharkiv** *Source*: http://en.wikipedia.org/w/index.php?title=University_of_Kharkiv *Contributors*: AdaDLHaji, Ahonc, Aldekein, AlexPU, Altenmann, D6, DDima, Daniel, Eduinukraine, Edward321, Elmondo21st, Epeefleche, Fartherred, Greenshed, Grosssmeister, Gun Powder Ma, Hgtp, Irpen, JamesBWatson, JamieS93, KPbIC, Khazar, Leutha, Mzajac, Nick UA, Noted Seven, Ornil, Patoldanga'r Tenida, Pinots, Qclam, Silin2005, Singhhardeep007, Skier Dude, Swarm, Unomano, Victor Korniyenko, WATERSLIDE, Waacstats, White guardian, Whiteroll, Васьйа, 49 anonymous edits

**Lev Landau** *Source*: http://en.wikipedia.org/w/index.php?title=Lev_Landau *Contributors*: 1812ahill, Abune, Afaber012, Alan Peakall, Alex Bakharev, Altenmann, Alvez3, Amillar, Ammarsakaji190, Ancheta Wis, Andres, AndriyK, Andrwsc, Aris Katsaris, Av0id3r, Avaya1, BD2412, Balcer, BigHairyBoris, Blainster, Brad7777, Brandmeister, Brandmeister (old), CMG, CYD, Chicheley, Clarityfiend, Cmapm, CommonsDelinker, CoolKid1993, Crum375, Curps, D6, DFRussia, Dana boomer, Daniel Mietchen, Danielfong, Davshul, Deville, Dougie monty,

Ebyabe, Ejconard, El C, Eliezg, Emerson7, Enchanter, Erkcan, Erkki Thuneberg, Everyking, ExRat, Ezhiki, Fastfission, Florincoter, Fulvius, GaeusOctavius, Galloping Ghost U of I, GcSwRhlc, Gcm, Gene Nygaard, Gene s, Ggorelik, Giftlite, Gilisa, Gnomz007, Goudzovski, Greyhood, Gulmammad, Headbomb, Hermitage17, Hgtp, Hillman, Humus sapiens, Ideyal, Illyukhina, Infovarius, Jackbars, JackofOz, Jauhienij, JillandJack, Jimmyeatskids, Jorgenev, Jspiegler, Jsqqq777, KNewman, Karlhufbauer, Kasparov, Kbdank71, Koavf, Kope, Ksnow, LarsMarius, LevKamensky, LiDaobing, Liilliil, Likebox, Lockley, Lumidek, MWaller, Mallodi, Masterpiece2000, Materialscientist, MattOates, Mattpickman, Maximus Rex, MessinaRagazza, Metzujan, Mic, Michael C Price, Michael Hardy, Mnmngb, Monedula, Morpheus & Momus, Nikolay Yeriomin, Noeckel, Olivier, Orloffm, Pallab1234, Parishan, Parkyere, Paul Pieniezny, Paulcardan, Petrukhina, Physman, Physnick, Pulu, RG2, RS1900, RSStockdale, Rglovejoy, Rich Farmbrough, Richiez, RickK, Rms125a@hotmail.com, RockMagnetist, Ronark, Russavia, Salih, Salsa Shark, Sanders muc, Sapphirain, Scott MacLean, SebastianHelm, SidP, Silin2005, Sj, Sk741, Snowolf, Splash, Srleffler, SteinbDJ, StewartMH, Tamfang, Taras danko, Theodorekon, Timrollpickering, TobiasS, TomyDuby, TorontoFever, Tristes tigres, Unyoyega, Valip, Vargenau, VasilievVV, Vasil', Vicki Rosenzweig, Victor-435, Vital303, Vuong Ngan Ha, Vyznev Xnebara, WillowW, XJamRastafire, XaosBits, Yawmoght, Zloyvolsheb, Фрашкард, דוד55, 187 anonymous edits

# Image Sources, Licenses and Contributors

**File:Chebotaryov-meiman.jpg** *Source*: http://en.wikipedia.org/w/index.php?title=File:Chebotaryov-meiman.jpg *License*: unknown *Contributors*: Lkitrossky, Sfan00 IMG

**File:Meiman dissidents.gif** *Source*: http://en.wikipedia.org/w/index.php?title=File:Meiman_dissidents.gif *License*: unknown *Contributors*: Lkitrossky, Sfan00 IMG

**File:Baranavichy pl lenina .JPG** *Source*: http://en.wikipedia.org/w/index.php?title=File:Baranavichy_pl_lenina_.JPG *License*: unknown *Contributors*: Ruslan Raviaka

**File:Coat of Arms of BaranaviČy, Belarus.png** *Source*: http://en.wikipedia.org/w/index.php?title=File:Coat_of_Arms_of_BaranaviČy,_Belarus.png *License*: unknown *Contributors*: EugeneZelenko, Lokal Profil, Red Winged Duck, Vadim Akopyan

**file:Belarus location map.svg** *Source*: http://en.wikipedia.org/w/index.php?title=File:Belarus_location_map.svg *License*: unknown *Contributors*: User:NordNordWest

**File:Red pog.svg** *Source*: http://en.wikipedia.org/w/index.php?title=File:Red_pog.svg *License*: unknown *Contributors*: Anomie

**File:Flag of Belarus.svg** *Source*: http://en.wikipedia.org/w/index.php?title=File:Flag_of_Belarus.svg *License*: unknown *Contributors*: User:Zscout370

**File:Baranovichi Law Institute.jpg** *Source*: http://en.wikipedia.org/w/index.php?title=File:Baranovichi_Law_Institute.jpg *License*: unknown *Contributors*: Igor Mostitsky

**File:Baranovichi Fountain at Central Square.jpg** *Source*: http://en.wikipedia.org/w/index.php?title=File:Baranovichi_Fountain_at_Central_Square.jpg *License*: unknown *Contributors*: Igor Mostitsky

**File:Belarus-Baranavichy-Ballistic Missile Monument.jpg** *Source*: http://en.wikipedia.org/w/index.php?title=File:Belarus-Baranavichy-Ballistic_Missile_Monument.jpg *License*: unknown *Contributors*: Bdk, EugeneZelenko, KGyST, Leonidl, Vadim Akopyan, VargaA

**File:Speaker Icon.svg** *Source*: http://en.wikipedia.org/w/index.php?title=File:Speaker_Icon.svg *License*: unknown *Contributors*: Blast, G.Hagedorn, Mobius, 2 anonymous edits

**File:Flag of Poland.svg** *Source*: http://en.wikipedia.org/w/index.php?title=File:Flag_of_Poland.svg *License*: unknown *Contributors*: Anomie, Mifter

**File:Flag of Italy.svg** *Source*: http://en.wikipedia.org/w/index.php?title=File:Flag_of_Italy.svg *License*: unknown *Contributors*: Anomie

**File:Flag of Finland.svg** *Source*: http://en.wikipedia.org/w/index.php?title=File:Flag_of_Finland.svg *License*: unknown *Contributors*: User:SKopp

**File:Flag of Latvia.svg** *Source*: http://en.wikipedia.org/w/index.php?title=File:Flag_of_Latvia.svg *License*: unknown *Contributors*: User:SKopp

**File:Flag of Bulgaria.svg** *Source*: http://en.wikipedia.org/w/index.php?title=File:Flag_of_Bulgaria.svg *License*: unknown *Contributors*: User:SKopp

**File:Flag of Russia.svg** *Source*: http://en.wikipedia.org/w/index.php?title=File:Flag_of_Russia.svg *License*: unknown *Contributors*: Anomie

**File:Flag of Ukraine.svg** *Source*: http://en.wikipedia.org/w/index.php?title=File:Flag_of_Ukraine.svg *License*: unknown *Contributors*: User:Jon Harald Søby, User:Zscout370

**File:Flag of Austria.svg** *Source*: http://en.wikipedia.org/w/index.php?title=File:Flag_of_Austria.svg *License*: unknown *Contributors*: User:SKopp

**File:Domenico-Fetti Archimedes 1620.jpg** *Source*: http://en.wikipedia.org/w/index.php?title=File:Domenico-Fetti_Archimedes_1620.jpg *License*: unknown *Contributors*: A. Wagner, Andreagrossmann, AndreasPraefcke, Bukk, Christophe.Finot, FranzK, Gene.arboit, Ianmacm, Kilom691, Kramer Associates, Luestling, Mattes, Plindenbaum, Serge Lachinov, Shakko, Wst, 3 anonymous edits

**File:GodfreyKneller-IsaacNewton-1689.jpg** *Source*: http://en.wikipedia.org/w/index.php?title=File:GodfreyKneller-IsaacNewton-1689.jpg *License*: unknown *Contributors*: Algorithme, Beyond My Ken, Bjankuloski06en, Grenavitar, Infrogmation, Kelson, Kilom691, Porao, Saperaud, Semnoz, Siebrand, Sparkit, Thomas Gun, Wknight94, Wst, Zaphod, 4 anonymous edits

**File:Leonhard Euler.jpg** *Source*: http://en.wikipedia.org/w/index.php?title=File:Leonhard_Euler.jpg *License*: unknown *Contributors*: User:Wars

**File:Carl Friedrich Gauss.jpg** *Source*: http://en.wikipedia.org/w/index.php?title=File:Carl_Friedrich_Gauss.jpg *License*: unknown *Contributors*: Bcrowell, Blösöf, Conscious, Gabor, Joanjoc, Kaganer, Kilom691, Luestling, Mattes, Rovnet, Schaengel89, Ufudu, Wolfmann, 4 anonymous edits

**File:JH_Poincare.jpg** *Source*: http://en.wikipedia.org/w/index.php?title=File:JH_Poincare.jpg *License*: unknown *Contributors*: Dabomb87, Mdd, Хацкер

**File:Hilbert.jpg** *Source*: http://en.wikipedia.org/w/index.php?title=File:Hilbert.jpg *License*: unknown *Contributors*: DaTroll, Darapti, Der Eberswalder, Gene.arboit, Harp, Mschlindwein, Yann, Zwikki, 3 anonymous edits

**File:Noether.jpg** *Source*: http://en.wikipedia.org/w/index.php?title=File:Noether.jpg *License*: unknown *Contributors*: Anarkman, Awadewit, Darapti, Factumquintus, Gene.arboit, Lambiam, Lobo, Romary

**Image:Color complex plot.jpg** *Source*: http://en.wikipedia.org/w/index.php?title=File:Color_complex_plot.jpg *License*: unknown *Contributors*: User:Rocchini

**Image:Mandel zoom 00 mandelbrot set.jpg** *Source*: http://en.wikipedia.org/w/index.php?title=File:Mandel_zoom_00_mandelbrot_set.jpg *License*: unknown *Contributors*: User:Wolfgangbeyer

**File:Heat eqn.gif** *Source*: http://en.wikipedia.org/w/index.php?title=File:Heat_eqn.gif *License*: unknown *Contributors*: User:Oleg Alexandrov

**Image:Etalon-big.jpg** *Source*: http://en.wikipedia.org/w/index.php?title=File:Etalon-big.jpg *License*: unknown *Contributors*: Vasyl Karazin Kharkiv National University

**File:Kharkiv State University.jpg** *Source*: http://en.wikipedia.org/w/index.php?title=File:Kharkiv_State_University.jpg *License*: unknown *Contributors*: User:EvgenyGenkin

**File:Kharkiv University2.jpg** *Source*: http://en.wikipedia.org/w/index.php?title=File:Kharkiv_University2.jpg *License*: unknown *Contributors*: User:NickK

**File:Lev Davidovich Landau.jpg** *Source*: http://en.wikipedia.org/w/index.php?title=File:Lev_Davidovich_Landau.jpg *License*: unknown *Contributors*: User:Mdd4696

**File:Landau1910.jpg** *Source*: http://en.wikipedia.org/w/index.php?title=File:Landau1910.jpg *License*: unknown *Contributors*: Фрашкард

**File:Landau Baku-cropped.JPG** *Source*: http://en.wikipedia.org/w/index.php?title=File:Landau_Baku-cropped.JPG *License*: unknown *Contributors*: User:Ds02006

Printed by Books on Demand GmbH, Norderstedt / Germany